Medical Physics for Veterinary and Related Studies: An Introductory Textbook on Mathematical and Physical Principles

Mehmet Erman Or

Medical Physics for Veterinary and Related Studies: An Introductory Textbook on Mathematical and Physical Principles

 Springer

Mehmet Erman Or
Faculty of Veterinary Medicine
Istanbul University - Cerrahpasa
Istanbul, Türkiye

ISBN 978-3-031-97354-3   ISBN 978-3-031-97355-0   (eBook)
https://doi.org/10.1007/978-3-031-97355-0

This Springer imprint is published by the registered company Springer Nature Switzerland AG
The registered company address is: Gewerbestrasse 11, 6330 Cham, Switzerland

If disposing of this product, please recycle the paper.

To my mother and father

Preface

The Medical Physics course, which I do not remember which faculty member taught it during my student years, included many topics that were not related to veterinary medicine that I learned during my high school education. The faculty member teaching the course changes every year and since he came from another faculty, it was not possible for us to find him at our faculty outside of class hours. I believe that this course, which I started to teach, will both enlighten students on how to use some basic information and methods in the field of medicine and prepare them for the clinical courses that will be taken in higher grades. Naturally, in order not to disrupt the integrity of some topics, various sections that are not related to veterinary medicine are also included in this book. In addition, some mathematical principles have been given in advance regarding some formulas that will be used in a later section during the course. In addition, questions related to topics that the student cannot find in any other source as a whole have been given at the end of the sections.

I believe that this book, which will be published with the approval of the internationally renowned publishing house *Springer Nature*, will be a first in the field of Veterinary Medicine and will be a source of pride for both me and my university. I also remember with gratitude my late Mathematics teacher Burçin ÜSTÜNKAYA and my late teacher Prof. Dr. Çetinkaya ŞENDİL, who believed at that time that I could convey this information in the future.

I hope this book will be useful to those who read it. I would like to thank Said CAFER, a PhD student who is an example of diligence and discipline, for helping me during the rewriting of the book; Asst. Prof. Dr. Bengü BİLGİÇ, who already shows signs of becoming a very good teacher in the future; and my dear wife Esra DİLER OR, who has always supported me with understanding.

İstanbul, Turkey   Mehmet Erman Or
29 March 2025

Contents

Introduction to Medical Physics

Abstract

In this section, the definition of science is made, and it is emphasized that various scientists work in more than one field, not in a single field, and the relationship between Mathematics, Physics, and Chemistry and Biology is emphasized. Basic concepts and derived concepts are specified; their definitions, symbols, formulas, and units in various systems are exemplified; and concept analyses and unit analyses are performed. Information is given about the system theory that is included in all fields of science, and biological examples are made by establishing its relationship with cybernetics. The feedback mechanism is evaluated especially over hormones. Both analytical geometry and biological analyses are made over directly proportional and inversely proportional quantities. By establishing a relationship between blood parameters and proportional quantities, blood collection locations in different animal species are specified, and anemia typing and leukocyte differences are examined with some special case questions. Biological relationships are established between derivatives and integrals in mathematics and trigonometric and logarithmic functions, and especially the relationship between the pH concept and the exponential function in X-ray physics is evaluated. At the end of each section, the subject is detailed with explanatory questions.

Keywords

Basic concepts · Derived concepts · Feedback mechanism · Blood parameters

M. E. Or, *Medical Physics for Veterinary and Related Studies: An Introductory Textbook on Mathematical and Physical Principles*,
https://doi.org/10.1007/978-3-031-97355-0_1

1.1   Subject of Medical Physics: History

Medical Physics, or *Biophysics* in a more general sense, is a discipline that deals with the use of experimental methods of physics in the study of living organisms. It attempts to explain events with the concepts, principles, and laws of physics. Biophysics can never be limited to using physical tools such as microscopes, electrocardiographs, ultrasonographs, electromyelographs, electroencephalographs, and endoscopes for various purposes. Biophysics, as an integrative discipline, requires specialized knowledge of many branches of science. For example, in order to understand all the stages of a vision event, knowledge of many special fields such as geometry, optics, spectroscopy, quantum physics and chemistry, physiology, psychology, and electronics is required.

From the perspective of biophysics, living organisms are open, self-regulating, productive, developing, heterogeneous, and very complex systems. Every stage of life, from small molecules to large biopolymers (proteins and nucleic acids), cell organelles, cells, tissues, organs, organisms, societies, and the biosphere, is also of interest to biophysics. Biophysics can be divided into three main branches with no clear boundaries between them:

(a) **Molecular Biophysics:** It deals with the physicochemical properties of biomolecules and especially biopolymers that have important functions.
(b) **Cell Biophysics:** It deals with the structures and functions of cells and tissues. The subjects that cell biophysics deals with include bioenergetic processes, biological membranes, nerve conduction, mechanochemical processes, photochemical events, etc.
(c) **Systems Biophysics:** It tries to explain the functioning of physiological systems with physical and mathematical methods and models.

Although until the end of the nineteenth century science was divided into branches such as astronomy, mathematics, physics, chemistry, biology, and psychology, scientists showed their creativity in various fields with a multidisciplinary perspective. Most of these people also received medical training.

The famous Turkish scholar Ibn-i Sina (980–1037) worked in various scientific fields along with medicine and pharmacy. Italian painter and sculptor Leonardo Da Vinci (1452–1519) was also one of the first to draw an anatomy atlas. Galileo Galilei (1564–1642) became a professor of mathematics after his medical education and worked on astronomy. In addition to his exploratory physics studies, he discovered that the period of a lamp in a cathedral across from where he was staying was independent of its amplitude by using his own pulse as a clock. After Galileo, Sanctorius (1561–1636) was the first researcher to introduce the pendulum and thermometer into clinical practice and to initiate quantitative measurements in medicine. William Harvey (1578–1657) was the first to use mathematical techniques in biological research and established the theory of blood circulation. The famous English mathematician Brook Taylor (1685–1731) conducted research on the determination of oscillation, percussion, and curvature centers and vibrating strings and experimentally examined the scientific problems of his time. The Scottish mathematician Colin MacLaurin (1698–1746), who lived in the same period, made new additions to Taylor's views with his book on fluids. The physician Luigi Galvani (1737–1798) observed contractions in frog muscles with two different metals and was the first scientist to be interested in bioelectric events. Thomas Young (1773–1829) was a professor of physics as well as a physician, developed a theory on color vision, and measured cell diameters in the blood using the wave theory of light he developed. He also conducted studies on elasticity and was interested in the fluidity of blood vessels. The French physician Jean Leonard Marie Poiseuille (1797–1869) conducted research on viscous flow laws and especially ensured that the units of viscosity were named after him. German physiologist Adolf Fick (1829–1901) worked on the laws of diffusion and developed techniques for measuring blood flow and volume. Surgeon Julius Robert Mayer (1814–1878) first used the principle of conservation of energy in biological systems by using the relationship between heat, work, and physiological processes. Hermann Ludwig Ferdinand von Helmholtz (1821–1894) measured the speed of conduction in nerves by examining muscle contractions, invented the ophthalmoscope (an instrument with different apparatus for examining the eye and ear; it is called an otoscope for the ear and an ophthalmoscope for the eye), and conducted research on color vision and hearing.

The founder of the Republic of Turkey, the Great Leader Atatürk (1881–1938), in addition to his military and political genius, wrote a geometry book with his own hand in the Dolmabahçe Palace in the winter of 1937 and contributed personally to the derivation and introduction into the Turkish field of application of many concepts still used today (dimension, space, surface, diameter, radius, section, arc, circle, tangent, angle, bisector, alternate interior angle, alternate exterior angle, base, horizontal, vertical, perpendicular, correspondent, position, triangle, quadrilateral, pentagon, diagonal, equilateral, isosceles, parallelogram, lateral, trapezoid, plus, minus, multiplication, division, equal, total, ratio, proportion, derivative, area, assumption, justification).

1.2   Scientific Method and Basic Principles

Science is the whole of the processes carried out to find facts, organize information about them, and develop new theories. *Scientific endeavor* is an activity aimed at finding the order between our observations of the behavior of nature, the natural laws valid in nature, and the basic origin of these laws. Therefore, it is still debated whether science is an artistic concept or a discipline. In his book titled *Face Our Democracy* in 2007, Emre Kongar defined science as the art of asking the right questions and using the right methods to reach their answers.

1.2.1 Concepts

In order to understand and explain the properties of matter and the behavior of natural systems, some expression tools called *concepts* are needed. For example, speed, mass, acceleration, energy, density, cell, acid, and base concepts determine the quantities and qualities of nature. Some of these are *basic concepts* (mass, length, time, temperature, current intensity, light intensity, and amount of matter) and some are *derived concepts* that can be defined with the help of basic concepts (surface, volume, density, acceleration, force, energy-work, electric charge, illumination, moment, viscosity, etc.).

It is possible to express the physical quantities corresponding to some concepts with a measurement number and unit. For example, the mass of an object can be expressed with a measurement number and unit. These types of quantities are called *scalar quantities*. For example, for mass,

$$m_1 = 9 \text{ kg}$$

For some concepts, only the measurement number and unit are not enough. For example, when a horse is running at 50 km/h, it is not fully expressed. It is also necessary to specify in which line and direction the horse is running. These types of concepts that need to be specified with line, direction, and starting point in addition to the measurement number and unit are called *vector concepts*, for example, force, speed, and acceleration.

Directed line segments are called *vectors*. A vector is determined by four parameters (Fig. 1.1).

1.2.2 Symbols

These are letters or signs that correspond to all concepts used in different branches of science and are used without change all over the world. In this way, a common language unity has been established between various disciplines and languages. While some of these symbols are used for only one concept, sometimes the same symbol corresponds to different concepts. In addition, showing some symbols with uppercase or lowercase letters expresses completely different concepts. The symbols for various concepts are given below:

V (speed), (volume), (electrical potential)
a (acceleration)
F (force)
P (pressure), (power)
d (density), (distance)

Fig. 1.1 Parameters representing the vector

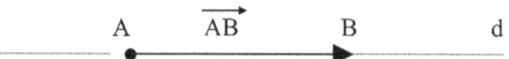

t (time), (temperature)
h (height)
x (distance)
m (mass), (magnetic pole strength), (apparent brightness)
M (moment), (absolute brightness)
T (tensile force), (period), (absolute temperature)
G (weight)
g (gravitational acceleration)
W (work)
w (angular velocity)
Q (heat)
q (electric charge)
E (illumination), (electromotive force)
Φ (luminous flux), (magnetic flux)
I (luminous intensity), (current intensity)
f (frequency), (focus)
n (refractive index)
D (angle of deviation)
c (length of the object)
g (length of the image)
u (distance of the object)
u' (distance of the image)
Y (convergence)
λ (wavelength)
Δ (diffraction)
C (electrical capacitance)
R (resistance)
ρ (resistivity)
r (internal resistance), (radius)
L (luminosity-radiative power)
A (absorption)
η (viscosity)

1.2.3   Formulas

In order to define and use derived concepts, different formulas belonging to the relevant fields have been used. Formulas can be created with the help of the definitions of these concepts, and a new concept can be defined by using the given formulas. For example, density is the mass of a unit volume of a substance. It can also be defined as the amount of 1 cm^3 in grams and its formula is expressed as d = m/V. In fact, the units and dimensional analyses of derived concepts are always done with the help of these formulas. The definitions and formulas of some derived concepts are given below:

Volume: The space occupied by a substance

The volume formulas for various geometric objects are generally accepted as base area × height (cube, rectangular prism, cylinder). In some geometric objects (pyramid, cone), this formula is applied as base area × height/3. For the sphere, the formula $4/3\ \pi r^3$ is used specifically.

Speed: Change in location per unit of time

$$v = \Delta x / \Delta t$$

Acceleration: Change in velocity per unit of time

$$a = \Delta v / \Delta t$$

Force: An influence that changes the shape and motion of an object

$$F = m.a$$

Weight: The gravitational force acting on an object

$$G = mg$$

Work: If an object changes its place under the influence of a certain force, it does work and consumes energy.

$$W = F\ \Delta x$$

Energy: Ability to do work

$$E = W$$

Power: Work done per unit of time

$$P = W / t$$

Pressure (solid): Force acting perpendicularly to a unit surface area

$$P = F / S$$

Pressure (liquid): The effect exerted by a liquid in a container on every point of the container due to its weight and the movement of liquid molecules

$$P = h.d.g$$

Viscosity: The force that creates a speed difference of 1 cm/s between the upper and lower surfaces of a cubic liquid mass that is 1 cm thick and has a surface area of 1 cm^2

$$\eta = Fd / Av$$

Moment: The rotational effect of a force

$$M = F.d$$

Frequency: The number of vibrations occurring per unit time

$$f = 1/T$$

Period: The time it takes for one complete vibration to occur

$$T = 1/f$$

1.2.4 Unites

To measure a quantity, first a unit size for that quantity is selected. However, since it is impossible to select a unit size for every quantity used in physics, basic units have been defined. The most common unit systems encountered in scientific publications are the CGS (length (cm), mass (g), time (s)) and MKS (length (m), mass (kg), time (s)) unit systems.

In recent years, the CGS unit system has also begun to be abandoned, and the scope of the MKS unit system has been expanded. Accordingly, by adding temperature (Kelvin), radiation intensity (Candela), current intensity (Ampere), and amount of substance (mole) to the basic units, a unit system that can be used in common by all branches of science has been obtained. The new system is called the *International System of Units*, or *SI* units for short. The units for basic concepts for CGS and SI are given in Table 1.1.

All other units are derived from the basic units, some of which are specifically named differently. Table 1.2 is designed specifically to take the SI unit system into account.

Apart from these, there are also units that are used together with SI units or can be converted to the CGS unit system (Table 1.3).

There are also some prefixes used with these units. These suffixes, their symbols, and conversion factors are given in Table 1.4.

Table 1.1 Basic concepts and their equivalents in the international unit system and the CGS unit system

	Symbol	CGS	MKS-SI
Length	x	cm	m
Mass	m	g	kg
Time	t	s	s
Heat	T		K (Kelvin)
Current intensity	I		A (Ampere)
Light intensity	I		cd (Candela)
Amount of substance	N		Mol

Table 1.2 Symbols and units of various derived concepts

Concept	Symbol	Unit	Special unit
Surface (area)	S (A)	m^2	
Volume	V	m^3	
Speed	V	m/s	
Acceleration	a	m/s^2	
Density	d	kg/m^3	
Force	F	kgm/s^2	(Newton=N)
Work, energy, heat	W	kgm^2/s^2	(Joule=J)
Power	P	kgm^2/s^3	(Watt=W)
Frequency	ν	1/s	(Hertz=Hz)
Pressure	P	kg/ms^2	(Pascal=Pa)
Electric charge	q		(Coulomb=C)
Electrical potential	V		(Volt=V)
Electrical resistance	R		(Ohm=Ω)
Electrical capacitance	C		(Farad=F)
Magnetic flux	Φ		(Weber=Wb)
Radiant flux	Φ		(Lumens=lm)
Illumination	E		(lux=lx)
Activity	R		(becquerel=Bq)
Viscosity	η		(Poiseuilli=PL)

Table 1.3 Different concepts and units

Concept	Unit name	Symbol	Conversion	Description
Time	Minute		60 s	
	Hour		3600 s	(60 min)
	Day		86400 s	(24 h)
Volume	Liter	l	$1 \ dm^3 = 10^{-3} \ m^3$	
Temperature	Celsius	°C	K	(T(K) = t (C) + 273)
Mass	Ton	t	1000 kg	
Pressure	Atmosphere	atm	76 cmHg = 760 mmH g = 760 Torr $1033.6 \ g/cm^2 = 1.0336 \ kg/cm^2$	
Activity	Curie	Ci	3.7×10^{10} Bq	
Dose	Roentgen	R	2.58×10^{-4} C/kg	
Length	Angstrom	A	10^{-10} m	
Force	Dyn	dyn	10^{-5} N	
Work, energy	Erg	erg	10^{-7} J	
Heat	Calorie	cal	4.18 J	
Pressure	Bar	bar	10^5 Pa	
Viscosity	Poise	P	$10 \ dyn.s.cm^{-2}$	

Table 1.4 Uppercase and lowercase prefixes

Multiplier	Prefix	Symbol
10^{12}	tera-	T
10^{9}	giga-	G
10^{6}	mega-	M
10^{3}	kilo-	k
10^{2}	hekto-	h
10	deka-	da
10^{-1}	desi-	d
10^{-2}	santi-	c
10^{-3}	mili-	m
10^{-6}	mikro-	μ
10^{-9}	nano-	n
10^{-12}	piko-	p
10^{-15}	femto-	f
10^{-18}	atto-	a

Table 1.5 Names, symbols, and values of constants

Constant	Symbol	Value
Speed of light	c	3×10^{10} cm/s $= 3 \times 10^{8}$ m/s
Planck constant	h	6.6×10^{-27} erg.s
Rydeberg constant	R	109677
Boltzmann constant	k	1.38×10^{-16} erg/K
Avogadro constant	N_0	6.02×10^{23} 1/mol
Stefan constant	σ	5.67×10^{-5} erg/cm^{2}sec.K^4
General gas constant	R	8.31 J.1/K.1/mol
Gravitational acceleration	g	9.81 m/s^2
Universal gravitational	G	$(6.673 \pm 0.003)\ 10^{-11}$ Nm^2/kg^2

For example, the unit of mean corpuscular volume, known as MCV, is the femtoliter.

1.2.5 Constants

These are constant expressions used in various formulas; the most commonly used ones are given in Table 1.5.

1.2.6 Dimensional Analysis

The dimensions of the expressions belonging to the basic concepts and the derived concepts obtained with the help of these concepts are given in Table 1.6.

Table 1.6 Various concepts and dimensions

	Dimension
Length	[L]
Mass	[M]
Time	[T]
Temperature	[K]
Current intensity	[A]
Light intensity	[I]
Amount of substance	[N]
Area	$[L^2]$
Volume	$[L^3]$
Density	$[ML^{-3}]$
Speed	$[LT^{-1}]$
Acceleration	$[LT^{-2}]$
Force	$[MLT^{-2}]$
Energy-Work	$[M\,L^2T^{-2}]$
Electric Charge	$[A\,T]$
Illumination	$[IL^{-2}]$
Moment	$[M\,L^2T^{-2}]$
Weight	$[MLT^{-2}]$
Power	$[M\,L^2T^{-3}]$
Pressure	$[M\,L^{-1}T^{-2}]$
Viscosity	$[M\,L^{-1}T^{-1}]$

Questions 1

1. Among domestic mammals, dogs have an average erythrocyte diameter of 0.007 mm while goats have an average diameter of 4,100,000 pm. In which animal species is the number of erythrocytes seen in a microscope field lower?

$$0.007 \text{ mm} = 7\ \mu m \quad (\text{Dog})$$
$$4.100.000 \text{ pm} = 4.1\ \mu m \ (\text{Goat})$$

Since the erythrocyte diameter and the number of erythrocytes seen in a microscope field are inversely proportional, more erythrocytes are observed in the goat sample.

2. Among domestic mammals, the average erythrocyte diameter of cats is 0.09 mm while that of deer is 3200 nm. In which animal species is the number of erythrocytes seen in a microscope field higher?

$$0.9 \text{ mm} = \text{......}\mu m \quad (\text{Cat})$$
$$3.200 \text{ nm} = \text{......}\mu m \ (\text{Deer})$$

Since the number of erythrocytes seen in a microscope field is proportional to .... and............., more erythrocytes are observed in the .... sample.

3. Between a Japanese Akita breed dog with an MCV value of 60 fl and a street dog with an MCV value of 70,000 pL, which one has a larger erythrocyte diameter?

Japanese Akita breed dog MCV value .... 60 fl

Street dog MCV value ....70,000 pl = ... fl

When the values are compared after conversion, it is revealed that the .... dog has larger erythrocytes.

When we compare the PCV values of these dogs when they are healthy, the ones with a large PCV value have a large MCV value when the RBC values are constant.

4. Prove that $1N=10^5$. The unit given as Newton is a force unit for SI. First, we need to reach the derived concept of force with the help of basic concepts.

5. Prove that $1 PL = 10 P$.

6. A scientist defined the concept of effort in biology as the power spent per unit of time. According to this definition, find the unit of effort using the basic concepts and units of the international unit system (SI) and convert it to its equivalent in CGS.

7. The elasticity of any solid body can be defined as pressure/strain. Strain is the ratio of the increase in the body's length to its initial length. According to these definitions, find the unit of elasticity using the basic concepts and units of the International System of Units (SI) and convert it to its equivalent in CGS.

8. A scientist defined concentration as the product of power and force per unit of time. According to these definitions, find the unit of concentration in the international unit system.

1.2.7 System and Behavior Equation

The General Systems Theory, proposed by biophysicist Ludwig von Bertalanffy (1920–1972), is an interdisciplinary mathematical field of study that aims to find and develop general principles and principles that can be applied to all types of systems. In other words, the first goal of the General Systems Theory was to develop tenet, principles, and theories that are a combination of branches of science such as Biology, Mathematics, Physiology, and Economics and that can be applied especially to issues such as growth and development. The systems approach is not a new scientific discipline in itself but rather a way of thinking used in the examination of certain events, situations, and developments.

A system is a whole consisting of parts (subunits or subsystems) that have certain relationships both with each other and with the external environment. The most important feature of this approach is that each of the parts that make up the whole has its own functioning characteristics, but the effectiveness of each is also dependent on each other. In physics, every whole that can be distinguished from its environment and can be active as an individual can be considered as a system and can be examined. H atom, water molecule, amino acids, and proteins are physical systems. A microorganism or a multicellular living being considered as a biological system can be examined as an auto-controlled system. For example, if a living body is considered as a system that tries to achieve certain goals, the nervous system, circulatory system, and urinary system in the organism can be considered as subsystems. That is why it is necessary to understand these subsystems in order to understand the whole. The purpose of the whole can only be achieved if these subsystems achieve their goals. The achievement of these subsystems is also largely dependent on each other. A change in one of the subsystems will affect the others. The properties of a system are determined by the types and numbers of its parts and the interactions between them. A system may emerge with new properties that its individual elements do not possess. An element within the system may acquire a positional value that it does not possess on its own.

Kenneth Boulding (1910) classified the systems on earth into nine groups in a specific hierarchy from simple to complex:

1. **Systems at the static structure level.** Examples include tables, chairs, and buildings.
2. **The level of simple dynamic systems with some specific movements.** Examples of this are the solar system, star systems, and the working of clocks.
3. **Control mechanism system or cybernetic system.** This system can automatically adjust itself to maintain balance. Examples of this are thermostats and machine guns.
4. **An open system that is self-protective and interacts with the environment.** We can show living cells as an example.
5. **Generic-social level system.** This system interacts with its environment. However, it is not mobile. We can show plants as an example.
6. **Animal system.** This system has an increasing mobility as it interacts with its environment. It is aware of itself; in other words, it searches for food to survive, escapes from dangers, and takes refuge in those it knows as friends.
7. **Human system.** This system can be effective outside the boundaries of its physical environment by interacting with the environment, using mobility, self-awareness, language and symbols.
8. **Human organization systems or social systems.** These are systems formed by groups of people, whether formally established or informally formed spontaneously, such as family, army, nation, state, school, business, and group of friends. What holds and integrates all people together are their common goals, languages, value and belief systems, and material and spiritual interests. These people find it appropriate to be together in a certain place and order on a certain day or at certain times of the day, not only in terms of interests but also emotionally. Because when a person is alone, he feels weak and powerless, and he wants to be together with other people; in other words, he is social. Every person, as well as being a system himself, enters social systems and becomes a part, element, or subsystem of it.

Fig. 1.2 Behavior equation

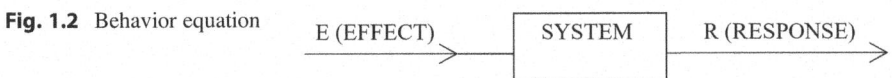

9. **Metaphysical systems.** These are the inevitable unknowns, the events whose causes cannot be fully explained. They reveal the systematic structure and relations. Like the postulates and relations in mathematics, it is not possible to prove them. However, their existence is accepted.

When any part of nature is considered as a system, the remaining parts constitute the environment of the system. Each system operates in a specific environment (surroundings). Everything that remains outside the boundaries of the system constitutes the environment. Systems that exchange matter and energy with their surroundings are called open systems, systems that are prohibited from exchanging matter are called closed systems, and systems that are prohibited from exchanging matter and energy are called isolated systems. In general, a system is schematized as follows (Fig. 1.2):

The energy or matter entering the system from the environment is called effect (E), and the energy released by the system to its environment is called response (R). Every system responds to the effect applied from the environment. In some branches of science, the terms input and output are used instead of the terms effect and response. The functional relationship between the response and effect is expressed as follows:

$$R = R(E, a, t) \text{ or } R = f(E, a, t)$$

Here, a is a structural parameter of the system, and t indicates time. Every natural event occurs according to a certain law. The law of nature is expressed in science as a relationship between concepts. The relationship that can be established between concepts through observation and experimentation is called the behavior equation of that physical system. The behavior equation is the determination of natural events that exist outside our consciousness with a certain approach. This basic learning has enabled our beliefs to change and superstitions to be destroyed.

Open systems constantly receive input from their environment and operate in a dynamic balance. They maintain dynamic balance by making changes in their internal structure according to changes in their environment. However, in closed systems, inputs are determined once and for all. Since this type of system has no relationship with its environment, it continues its activity until it stops. The importance of the environment comes from the fact that it has the potential to affect the system and its functioning. For example, if a dog that is used to living in a warm climate is forced to live in a very hot climate, this very hot environment will affect the system in various ways.

Another important concept in the system approach is the terms entropy and negative entropy (Negentropy). The concept of entropy, taken from thermodynamics, expresses a tendency in the whole that is called a system. In a system, there is a tendency for activities to deteriorate, balance to be lost, confusion and disruptions to appear, and eventually for the system to stop. Entropy is the concept that expresses this tendency. Therefore, there is entropy in all systems, regardless of their quality and size. In other words, there is a tendency for confusion, disorder, deterioration, stagnation, and eventually a complete stop in systems. Entropy is strong in closed systems and is the most important factor that stops the system after a certain period of time. However, entropy can be stopped in open systems. So open systems can stop entropy and make its effects negative with the information, energy, and material they receive from their environment. Therefore, there is negative entropy (Negentropy) in open systems. For a biological system, maximum entropy means death.

Another feature of open systems is the principle of equifinality. They can reach the same final state through different initial conditions and different paths. For example, growth is an equifinal event. Individuals of the same species can reach the same characteristic final size by starting from different initial sizes and encountering different obstacles. For example, if the length of an organism is represented by L, anabolism is proportional to the surface area of the organism and can be written as anabolism $= \alpha L^2$. Since catabolism is proportional to the mass or volume of the organism, catabolism $= \beta L^3$. Growth, in terms of mass increase, is expressed as $dm/dt = \alpha L^2 - \beta L^3 > 0$ where m represents mass and t represents time.

It is possible if $dm/dt = \alpha L^2 - \beta L^3 > 0$. Growth stops and reaches a steady state when $dm/dt = 0$. For the final length of the organism in steady state, $L_{last} = \alpha/\beta$, which is independent of the initial conditions and is determined only by the system parameters.

Growth or advancement processes in unit time are defined as progressive events in biology. Hypertrophy (mass growth), hyperplasia (volumetric growth), and metaplasia (proliferation of another cell group in the place where there is a cell group) are progressive events, for example, hypertrophic cardiomyopathy or intestinal metaplasia.

In all progressive events, there is a development in the form of $dm/dt > 0$ or $\alpha L^2 - \beta L^3 > 0$. When drawing unit time mass change graphs, an upward linear graph can be mentioned.

The processes of shrinkage or regression in unit time are defined as retrogressive events in biology. Atrophy (mass shrinkage), hypoplasia (volumetric shrinkage), and matter accumulations-matter degenerations or necrosis (cell death) are retrogressive events, for example, brain atrophy or bone necrosis. Hyaline degeneration and amyloid accumulation can be given as examples of matter accumulations or degenerations. Colliquation necrosis, calcification necrosis, and gangrene can be given as examples of necrosis types.

In all retrogressive events, there is a change in the form of $dm/dt < 0$ or $\alpha L^2 - \beta L^3 < 0$. However, when unit time mass change graphs are drawn, a downward linear graph can be mentioned. However, when necrosis occurs, the graph drawn must intersect with the horizontal axis because death has occurred.

Based on this information, the following types of problems are generally encountered in scientific activities:

9.1. Direct Problem: Problem of determining the response when the effect applied to the system from the environment and the law are known

9.2. Opposite Problem: Problem of finding the effect applied by the environment when the response of the system and the law are known

9.3. Inverse Problem: Problem of finding the system when the effect, response, and law are known

9.4. Inductive Problem: Problem of finding the law when the effect, system, and the answer are known

I. Is P (a) correct?

II. Is P (k + 1) true when P (k) is true for $\forall$ k $\in$ A?

For example, $\forall$ n $\geq$ 1

Let's prove that $1 + 2 + 3 + \ldots + n = n(n+1)/2$.

I. If $1 = 1(1+1)/2 = 1$ for n = 1   then   P (1) is true

II. For n = k $1 + 2 + 3 + \ldots + k = k(k+1)/2$ is true

Let $1 + 2 + 3 + \ldots + k + (k+1) = (k+1)(k+2)/2$ be true for $n = k+1$

$$1 + 2 + 3 + \ldots + k + (k+1) = k(k+1)/2 + (k+1)$$
$$= \left(k^2 + k + 2k + 2\right)/2$$
$$= \left(k^2 + 3k + 2\right)/2$$
$$= (k+1)(k+2)/2$$

Some subsystem symbols and transfer functions used in automatic control systems diagrams (Fig. 1.3):

Questions 2

1. $Y_1 = \int E\,dx$ $Y_2 = dE/dx$ $Y_3 = 2E$

   $E_1 = e^x,$ $E_2 = \ln x,$ $E_3 = 2^{-1}$

   Considering the effects and response functions given above, find the total response of the system.

2. $Y_1 = e^E$ $Y_2 = \int E\,dx$ $Y_3 = dE/dx$ $Y_4 = \sqrt{E}$

   $E_1 = e^x,$ $E_2 = x^2 + 4,$ $E_3 = \ln x$ $E_4 = x^2$

   Considering the effects and response functions given above, find the total response of the system.

3. If a hyperplastic liver has necrosed after a certain period of time, how should the mass change graph be drawn?

4. If a metaplastic spleen later becomes gangrenous, how should the mass change per unit time graph be drawn and state the formulas showing the relationship between anabolism and catabolism activities for each section.

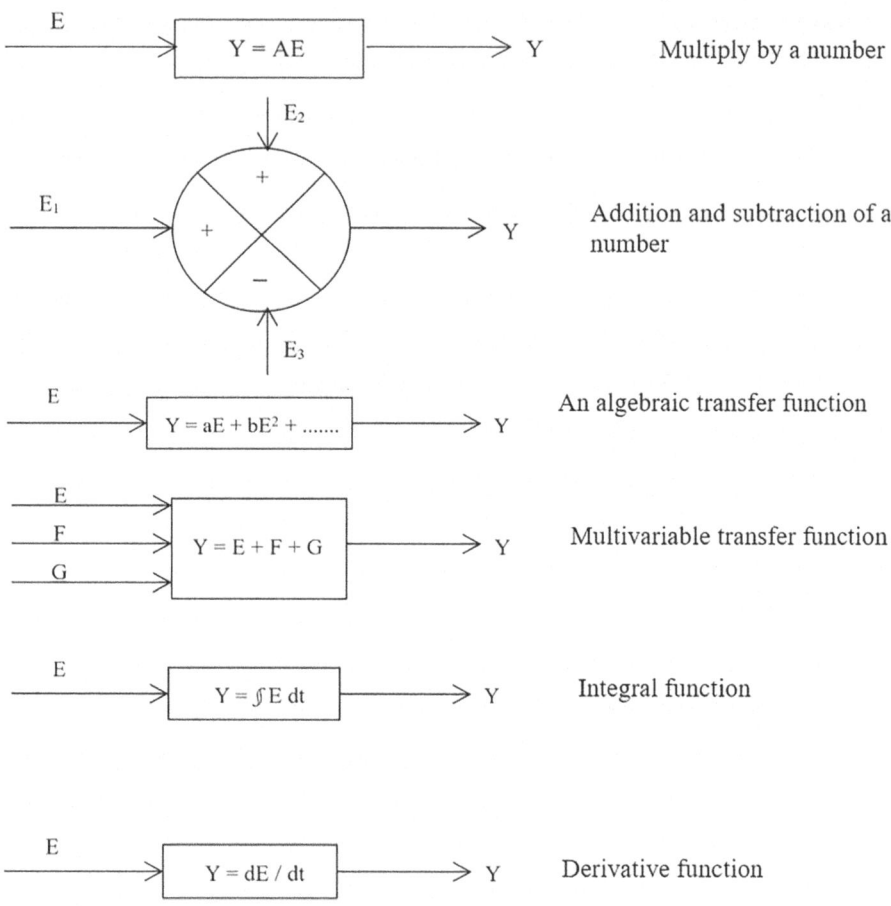

Fig. 1.3 Representations of various behavioral equations

1.2.8   Cybernetics

American mathematician Norbert Wiener (1894–1964) started to work on communication theory after 1940 due to the technical problems caused by World War II and thus laid the foundation of cybernetics. The researcher who explained his views in his book titled *Cybernetics: Or Control and Communication in the Animal and the Machine* determined great similarities in terms of control processes in living beings and machines that are completely different in structure. A branch of cybernetics called bionics is directed toward making machines (robots) that imitate human mental activities in light of the information about the functioning of the nervous system in humans, and another branch called biocybernetics is directed toward explaining the functioning of living beings, especially the nervous system, in light of the information about the functioning of electronic systems and machines. Logical Mathematics or Boolean algebra is used for this purpose. In Boolean algebra, there are two elements (0 and 1) and three operations (AND), (OR), and (NOT) as in the binary calculation system.

In logic mathematics, statements that are definitely true or false are called propositions. The true or false statements in this definition indicate the truth value of that proposition. For example, the statement "Nose clamp is a tool used to calm down slow horses" is a true proposition, while the statement "Ruminants have only one hoof" is a false proposition. However, the statement "What is its prognosis?" is not a proposition because it is a question. According to the truth value, false propositions are considered 0, and true propositions are considered 1. In addition, these propositions are represented in the electrical circuits shown in Figs. 1.4 and 1.5.

When this switch circuit is open, it indicates that no current will pass, that is, 0 state, and when it is closed, it indicates that current will pass, that is, 1 state.

The operation of a nerve cell in accordance with the all-or-nothing law is a cybernetic event that can be represented by this switch circuit and has only two possibilities, 0 or 1. The "Yes" response to a stimulus exceeding the threshold value of the nerve cell is expressed as 1, and the "No" response to a stimulus below the

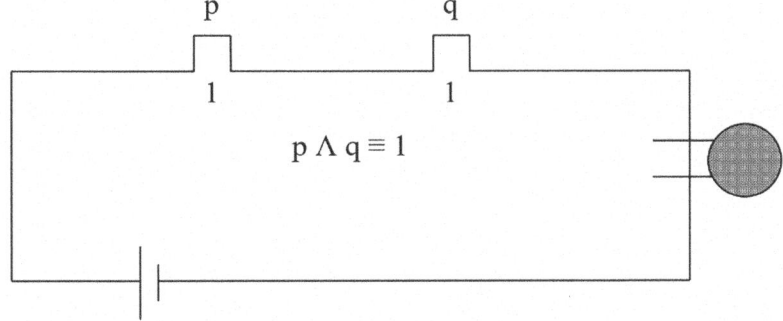

Fig. 1.4 Representation of the compound proposition And when both propositions are closed

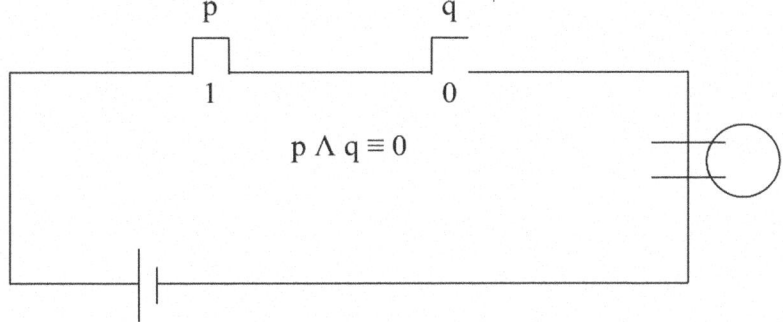

Fig. 1.5 Representation of the compound proposition And when one of the two propositions is closed and the other is open

threshold value is expressed as 0. A threshold value is mentioned for each nerve cell. If a message exceeds the threshold value, it is responded to by an electric current in the cell. If it does not exceed the threshold value, it is assumed to be absent and no response is received.

1. NOT (p′)
In the cybernetic model, they are expressions that have the inverse function of the key or the inverse truth value and are expressed as p′ or ~p.

2. OR (p V q)
It is the position related to the parallel connection of two switches in an electrical circuit (Fig. 1.6).

3. AND (p Λ q)
It is the position related to the series connection of two switches in an electrical circuit (Fig. 1.7).

If a nerve cell can be activated when the stimuli coming from the two synapses connected to it, indicated by x and y, reach the cell at the same time and overlap, then

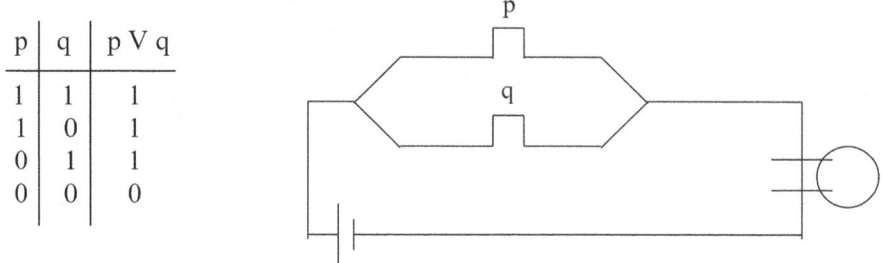

p	q	p V q
1	1	1
1	0	1
0	1	1
0	0	0

Fig. 1.6 Or representation as a compound proposition

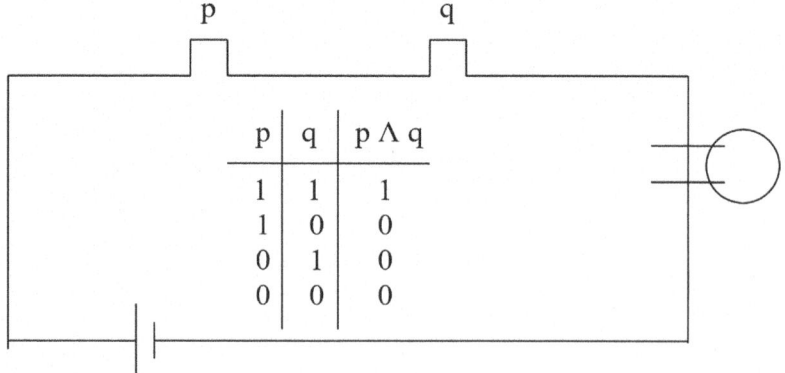

p	q	p Λ q
1	1	1
1	0	0
0	1	0
0	0	0

Fig. 1.7 And as a compound proposition

Table 1.7 Compound proposition If

p	q	p ⇒ q
1	1	1
1	0	0
0	1	1
0	0	1

$$p \Rightarrow q \equiv p' \lor q$$

$$(p \Rightarrow q)' \equiv p \land q'$$

Table 1.8 Only and only the combined proposition

p	q	p ⇔ q
1	1	1
1	0	0
0	1	0
0	0	1

$$p \Leftrightarrow q \equiv (p \Rightarrow q) \land (q \Rightarrow p)$$

the threshold value can be exceeded and it can be activated, then the AND situation is in question; if it can be activated by the stimulus coming from only one of the synapses, then the OR situation is in question.

There are also compound propositions if and if and only if (Tables 1.7 and 1.8).

4. THEN (p ⇒ q)

5. ONLY AND ONLY (p ⇔ q)

6. Tautology-Contradiction

Compound propositions whose truth value is definitely 1 for every situation are called tautologies, and compound propositions whose truth value is 0 are called contradictions. For example,

$$p \lor p' \equiv 1 \text{ (tautology)} \qquad p \land p' \equiv 0 \text{ (contradiction)}$$

$$p \lor 1 \equiv 1 \text{ (tautology)} \qquad p \land 0 \equiv 0 \text{ (contradiction)}$$

7. De Morgan Rules

I. $(p')' \equiv p$

II. $(p \lor q)' \equiv (p' \land q')$

III. $(p \land q)' \equiv (p' \lor q')$

In addition, similar to numbers in mathematics, there are also the properties of commutativity, associativity, and distribution. Unlike these, there is also the single power property defined as $p \lor p \equiv p$ and $p \land p \equiv p$.

8. Proof Methods

1. Direct proof: It is the method of showing that the judgment is correct by starting from the truth of the hypothesis.
2. Contradiction: It is a method of showing that the negative of the statement necessitates the negative of the hypothesis.

9. Quantifiers (Some-At Least One: ∃, Every-All, Whole: ∀)

In order for a proposition made with some quantifiers to be true, it is sufficient to give at least one example that proves the proposition. In order for a proposition made with every quantifier to be true, it is necessary not to be able to give even one example that does not prove the proposition.

$\exists x, x^2 - 1 = 0, x \in R$ The proposition is true
$\exists x, x^2 < 0, x \in R$ The proposition is false
Some birds don't fly   The proposition is true

$\forall x, -x^2 \leq 0, x \in R$ The proposition is true
Every number divides itself   The proposition is false
Every bird flies   The proposition is false

10. Binary System

This type of calculation called Binary System or Digital System is very slow but never makes mistakes because it is similar to finger counting. The abacus, which we can call the simplest calculator, is a mechanism consisting of balls placed on rows of wires and moving right and left, which is used by children who are just learning to calculate even today. A ball on the left does not print a number. That is, it indicates a no answer and a zero state. Pulling the ball to the right shows a yes answer; that is, it is a 1 state. If the top row shows 1s, the bottom shows 10s, and the bottom shows 100s, pulling a ball from the top prints 1, pulling 3 balls from the middle prints 30, and pulling 5 balls from the bottom prints 500, and the total is 531. It is possible to perform mathematical operations by adding or subtracting balls to this (Fig. 1.8).

Since every cell in our nervous system (each nerve cell can be likened to a ball on an abacus) is created according to the answer "yes" or "no," this error-free binary calculation system is also valid in our brain. Just as the calculation ability can be increased by increasing the number of balls in the abacus, the ability of the system is increased by increasing the number of cells that come into play in the nervous system.

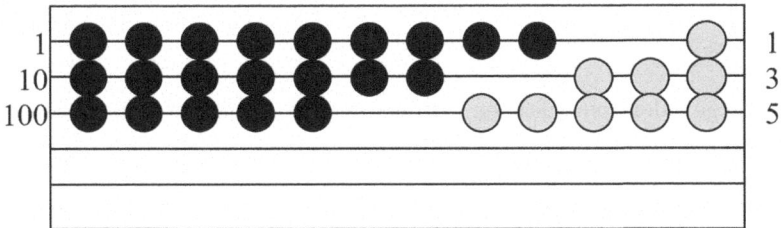

Fig. 1.8 Representation of the number 531 on the abacus as an example

11. Feedback (Backward Information Flow)

The event of reconnecting the result of a system's operation to the cause that caused the result and thus subjecting the result to control and adjustment is called "feedback." One of the best examples of this event is the regulator mechanism in a steam engine. When a steam engine is operating, it rotates a shaft around its own axis. Two heavy metal spheres are hung around this shaft. These spheres open sideways with centrifugal force, and this opening reduces the pressure of the incoming steam by reducing the steam cock; therefore, it reduces the rotation speed. The decrease in speed this time allows the spheres to close toward the axis, that is, the steam valve to open. In this way, the rotation speed of the machine will be kept constant within certain limits, and the load or the ups and downs on the road will not affect the speed of a locomotive, for example.

All biological systems contain feedback mechanisms. The opening and closing of the pupil according to the light intensity; the regulation of breathing rate and depth, heart rate, and blood pressure according to the oxygen needs of the tissues; and the regulation of blood sugar and body temperature are all possible by connecting the result back to the cause, that is, through feedback systems.

There are two types of feedback. If the result is always connected in a way that it affects the cause in the opposite direction, this mechanism is called negative feedback. A system that cools a heated system and heats a cooled system, thus keeping the temperature constant between certain degrees, is a negative feedback. Negative feedback refers to the flow of information that shows how much the system deviates from previously determined goals.

If the result is connected to the cause in a way that it has a directly proportional effect, this is called positive feedback. Such a connection eventually causes the destruction of the system. In some pathological events, such feedback controls are seen to occur; for example, blood sugar constantly increases, which ends with the death of the organism.

The most important subject in biology where the feedback system is observed is hormones. Hormones are natural chemical substances secreted from the endocrine glands that regulate functions by affecting body tissues or organs through the blood. We can classify hormones as follows:

1. GNRH (gonadotropin releasing hormone) secreted from the hypothalamus
2. Hormones secreted from the pituitary gland
   FSH: Follicle stimulating hormone
   LH: Luteinizing hormone
   ACTH: Adenocorticotropic hormone
   STH: Somatotropic hormone
   TSH: Thyroid stimulating hormone
   Prolactin
   Oxytocin
   ADH: Antidiuretic hormone, vasopressin
3. Hormones secreted from the gonads
   Estrogen
   Progesterone
   Androgen-testosterone
4. Hormones secreted from other endocrine organs
   Prostaglandins-uterus
   Glucocorticoids-adrenal glands
   Thyroid hormones

Negative Feedback
As seen in the diagram (Fig. 1.9), a large portion of hormones are secreted in an inductive manner from the center to the periphery on an axis that starts from the brain (cortical centers, limbic system including hypothalamus) and extends to the pituitary gland and from there to the peripheral hormone glands (adrenal gland, ovary, testis). When the concentration of the hormone secreted by the last hormone gland in the blood exceeds a certain level, this time the effect is reversed (negative feedback) and the secretion of the hormone that initiated the induction stops. For example, the estrous cycle begins with the sense of sight. This stimulus is transmitted to the hypothalamus via the optic nerve in the retina. From here, the production of the GnRH hormone begins; accordingly, with the effect of the FSH and LH hormones secreted by the pituitary gland, the follicle develops in the ovary and ovulation occurs (estrogens are secreted). Following ovulation, the corpus luteum transforms and the progesterone hormone is secreted.

While estrogens and progesterone prepare the female genital organs for fecundation and pregnancy, they also prevent a new estrus cycle by affecting the hypothalamus and pituitary gland via the bloodstream and stopping the secretion of GnRH, FSH, and LH hormones.

If pregnancy has occurred, the progesterone brake, or feedback effect, continues. If pregnancy has not occurred, the prostaglandin hormones formed in the uterus melt the corpus luteum. As a result, progesterone secretion stops. Since the brake on the

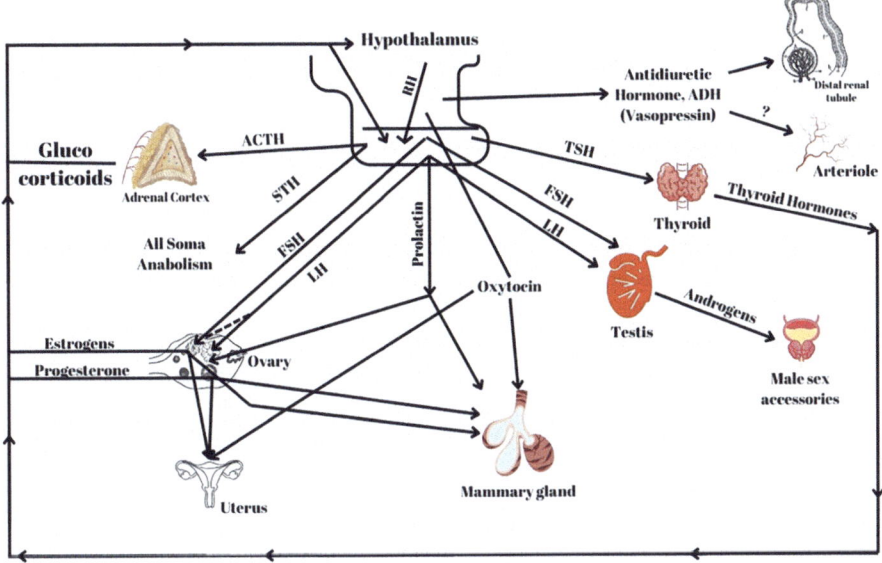

Fig. 1.9 Hormones and feedback mechanism

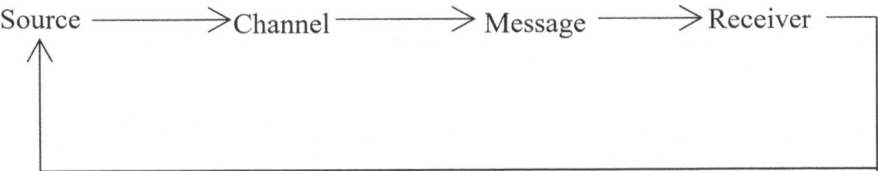

Fig. 1.10 Elements of communication

hypothalamus and pituitary gland is removed, a new cycle begins with GnRH secretion.

Similarly, there are mutual secretory interactions between estrogens and androgens and GnRH, FSH, and LH, between glucocorticoid hormones and CRH (corticotropin-releasing hormone) and ACTH hormone secretion, and between thyroxine and thyreotropic hormones.

The concept of communication can be examined as an example of the concept of feedback from a social perspective. Communication is defined as the process of sending and receiving meaning between two elements called the source and the receiver and influencing each other. There are four important elements in communication (Fig. 1.10):

After a receiver decodes a message and gives it meaning, it becomes a source. That is, it prepares a message as a sender to respond to the message it has received and transmits it to the old sender through the channel. This is called feedback in communication. Feedback provides the sender with information about whether the message was received or interpreted correctly.

Questions 3

1. If pregnancy occurs after the formation of the corpus luteum following ovulation, which hormone's effect continues? Where is this hormone secreted from?
   The effect of progesterone continues. This hormone is secreted from the ovaries.

2. If pregnancy does not occur after the formation of the corpus luteum following ovulation, which hormone is released from the uterus and which hormone is released again later with the feedback effect?
   Prostaglandins are secreted from the uterus and progesterone secretion stops. A new cycle begins with GNRH secretion.

3. Prove the statement $(q \Rightarrow p') \equiv (q' \wedge p)$ with the help of De Morgan rules and table.
   $$(q' \Rightarrow p')' \equiv (q \vee p')' \equiv q' \wedge p$$

q	p	q'
1	1	0
1	0	0
0	1	1
0	0	1

   $$p \Rightarrow q \equiv p' \vee q$$

   $$(p \Rightarrow q)' \equiv p \wedge q'$$

4. $[(\forall x, x - 1 \neq 0) \Rightarrow (\forall x, x - 2 > 0)]' \equiv ?$
   If we consider $(p \Rightarrow q)' \equiv p \wedge q'$
   $$[(\forall x, x - 1 \neq 0) \Rightarrow (\forall x, x - 2 < 0)]' \equiv [(\forall x, x - 1 \neq 0) \wedge (\exists x, x - 2 \geq 0)]$$

5. $[(\exists x, x - 1 \geq 0) \Rightarrow (\forall x, x - 2 = 0)]' \equiv ?$
   If we consider $(p \Rightarrow q)' \equiv p \wedge q'$
   $$[(\exists x, x - 1 \geq 0) \Rightarrow (\forall x, x - 2 = 0)]' \equiv [(\exists x, x - 1 \geq 0) \wedge (\exists x, x - 2 \neq 0)]$$

6. Given that $[(p \Rightarrow (q \vee r')] \equiv 0$, find whether the propositions p, q, and r are true or false.
   When $P \Rightarrow q \equiv 0$, $p \equiv 1$, $q \equiv 0$ must be true. Then the second proposition, that is,
   $(q \vee r') \equiv 0$ is true.
   If $p \vee q \equiv 0$, then both propositions must be 0. $q \equiv 0$ and $r' \equiv 0$, that is, $r \equiv 1$

7. Prove that $(p' \Rightarrow q)' \equiv p' \wedge q'$ using the table.

8. Since $[(\exists x, x: \text{hormones affect the testicles})] \equiv 1$, what can be written instead of x? For example, x = FSH or LH

9. If $[(\forall x, x: \text{hormones are released from the pituitary gland})] \equiv 0$, what can be written instead of x? For example, x = GNRH

10. A newly defined compound proposition named OR is shown with the symbol $\forall$ and gives false answers when the truth values of the propositions are the

same and true answers when they are different. Determine which of the following are true and which are false by making a table.

(a) $p \lor p \equiv 0$

(b) $p \lor p \equiv 1$

(c) $p \lor p \equiv p$

(d) $p \lor q \equiv q \lor p$

(e) $p \lor (q \lor r) \equiv (p \lor q) \lor r$

(f) $(p \lor q)' \equiv p' \land q$

11. Given that $(p \land q) \Rightarrow r) \equiv 0$, find the truth values of the expressions given below.

(a) $(p \land q) \land r$

(b) $r \lor (p \land q)$

(c) $r' \Rightarrow (p \lor q)$

(d) $r \Rightarrow (p \lor q)$

(e) $r' \land (p \lor q)$

12. Given that $p \Rightarrow (r' \lor q) \equiv 0$, find the truth values of the expressions given below.

(a) $p' \Rightarrow q$

(b) $p \Rightarrow (r \lor q)$

(c) $p \Rightarrow r'$

(d) $q \Rightarrow (p \lor r)$

(e) $(p \lor r)$

13. Given that $p' \Rightarrow (q \Rightarrow r) \equiv 0$, find the truth values of the expressions given below.

(a) $(p \lor q') \land r$

(b) $(p' \land q') \lor r$

(c) $(p \land q) \lor r'$

(d) $(p \lor q)' \lor r$

(e) $p \lor r$

14. $[(x - 3)(x + 2) \geq 0 \Rightarrow (x = 2 \lor x = 3)]' \equiv ?$

$[(x - 3)(x + 2)\ldots\ldots0\ldots\ldots(x\ldots\ldots2\ldots\ldots x\ldots\ldots3)]$

15. Identify the tautologies among the statements given below.

(a) $(p \lor q) \Rightarrow q$

(b) $p \Rightarrow (p' \land q)$

(c) $p' \Rightarrow q$

(d) $p' \Rightarrow (p \lor q')$

(e) $p' \Rightarrow (p \Rightarrow q)$

16. $(p' \lor q)' \lor (p \Rightarrow q')' \equiv ?$

$(p \land q') \lor (p \land q) \equiv p \land (q' \lor q) \equiv p \land 1 \equiv p$

17. $(a \land b) \Rightarrow a \equiv ?$

$(a' \lor b') \lor a \equiv a' \lor a \lor b \equiv 1 \lor b' \equiv 1$

18. $(p \Leftrightarrow q)' \equiv [(p \Rightarrow q) \land (q \Rightarrow p)]'$ Prove with the table.

19. $[(\forall x, x-1 > 9) \land (\exists x, x + 3 \neq x) \lor (\exists x, 2x + 3 \geq x)]' \equiv ?$

20. $[(\exists x, x + 1 \geq 0) \Rightarrow (\exists x, x + 1 \geq 3)]' \equiv ?$

21. Prove the statement a ≠ 3 ⇒ 5a-1 ≠ 14 by the method of proof by nullity.
    Hypothesis   Provision
        In this method, the negative of the verdict is expected to negate the
    negative of the hypothesis. Therefore,

$$5a - 1 = 14 \Rightarrow a = 3$$
$$5a \quad = 15$$
$$a \quad = 3$$

1.3 Important Mathematical Concepts in Terms of Medical Physics

1.3.1 Ratio-Proportion

The expression a/b, provided that both are not zero at the same time, is called a ratio. It is a concept that is used quite frequently in the field of medicine during the use of medicine. For example, a veterinarian who answers a patient owner who asks how to give milk to a new puppy they bought, by saying "mix it half with water," is using the concept of ratio. Again, the alcohol salicylate in the prescription written for a dog with eczema will be prepared by the pharmacist in the ratio requested by the veterinarian, with the addition of another drug if necessary.

The equality of two or more ratios is called proportionality. Proportionality is shown as a/b = c/d = k. k is called the constant of proportionality. One of the most important properties of the constant of proportionality is:

$$a/b = c/d = e/f = k \rightarrow (a + c + e)/(b + d + f) = k$$

The concept of proportion is also a concept that is constantly used, especially by doctors working in clinics. Continuous proportions are used in the dosage of drugs. For example, an antimycotic drug with the active ingredient ketoconazole is used in dogs at 10 mg/kg body weight. The doctor establishes a proportion when calculating the treatment dose according to the weight of the dog.

Questions 4

1. Dairy cows need more Mg because they lose 80–180 mg of Mg with 1 L of milk. If a dairy cow that produces 30 L of milk per day only benefits from 20% to 25% of the magnesium in the ration, how many grams of Mg should be in the daily ration at least?
   It loses an average of 130 mg of Mg with 1 L of milk. Therefore, with 30 L of milk, an average of
       $130 \times 30 = 3900$ mg of Mg = 3.9 g of Mg loss is in question.
       If dairy cows only benefit from 20% to 25% of the magnesium in the ration,
       $M \times 25/100 = 3.9$, and when the calculation is done, M = 15.6 g at least of Mg should be included in the daily ration.

2. The L-thyroxin active ingredient preparation used in the treatment of hypo-thyroidism is used in dogs at a dose of 0.002 0.004 mg/kg body weight. The total daily dose is divided into two. The commercial form of the drug contains 0.2 mg of active ingredient. Accordingly, how many tablets of the drug should be given to a dog weighing 30 kg in the morning?
3. There is 40 mg of active ingredient in 1 mL of drug A. For treatment, it is desired to give 10 mg of active ingredient per kg per day. How many liters of medicine should be given to a 20 kg dog?

1.3.2 Direct and Inverse Proportional Multipliers

1. Direct Proportion

If one of the quantities x and y increases while the other also increases, or if one decreases while the other also decreases, these two quantities are said to be directly proportional and are denoted by $y/x = k$. Directly proportional expressions are parameters in the form of parts. In other words, if two quantities are directly proportional, their ratios are constant.

If a graph is drawn according to certain values of directly proportional expressions, it becomes a linear graph. This graph can be expressed in several ways:

1. $y = mx$
   It is the equation of a straight line passing through the origin and having a slope of m.
2. $y = mx + n$
   It is a general line equation and represents lines that do not pass through the starting point.
3. $ax + by + c = 0$
   It is another expression of the general straight-line equation and can be transformed into the form $y = mx + n$ by rearranging. Here, $m = -a/b$ and $n = -c/b$.
4. $x/a + y/b = 1$
   It is known as the equation of the line that intersects the axes and can be easily converted to the general equation of the line. The parameters a and b must be used with their signs.

In fact, the relation $y = mx + n$ is used as a straight-line equation in analytical geometry (Fig. 1.11). However, when $n = 0$, the relation becomes $y = mx$. For example, if we examine how a dog's blood hemoglobin (HGB) value changes according to the hematocrit (PCV) value, HGB values are determined in response to PCV values known as percentages, and the HGB values in response to PCV values are plotted on the x and y Cartesian coordinate axes to show the change relationship graphically (Fig. 1.12).

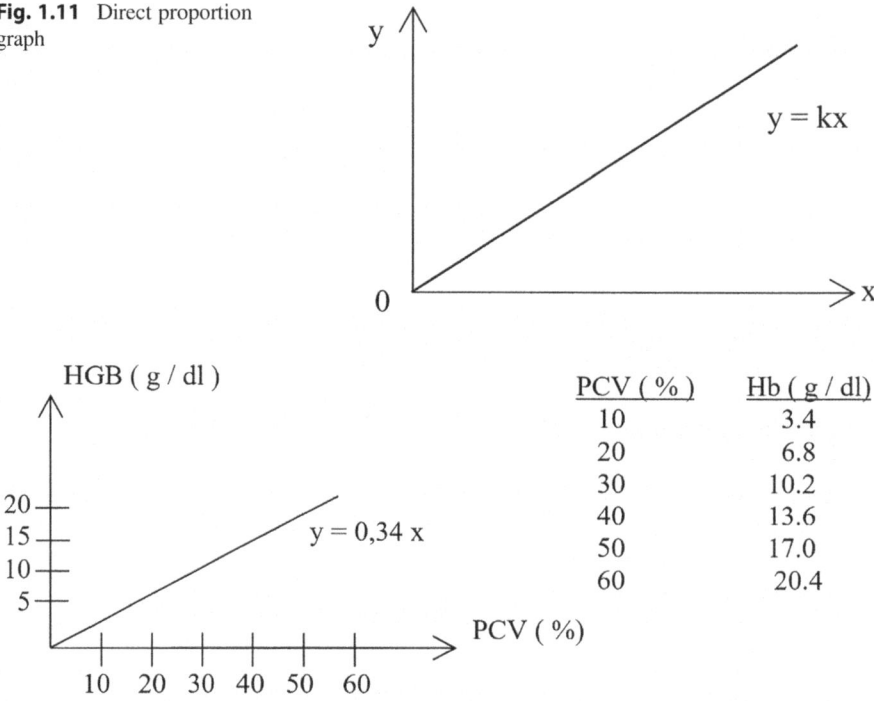

Fig. 1.11 Direct proportion graph

Fig. 1.12 Graph between hemoglobin and hematocrit

In the figure, the PCV value (%) is marked on the horizontal axis and the HGB value (grams per dL) is marked on the vertical axis. When the graph is examined, it is seen that the measurement points are arranged on a line passing through the origin. This shows us that there is a proportionality relationship between the HGB amount and the PCV value. The n parameter being zero means that hemoglobin is only found in the erythrocyte.

If $n = 0$ were not, then a linear relationship would be established in accordance with the relation $y = mx + n$. For example, let's examine how the plasma density of a living organism changes according to the total protein amount. Density determinations are made in response to total protein values known as grams per milliliter, and the density values in response to total protein amounts are plotted on the x and y axes in Cartesian coordinates and the change relationship is shown on the graph (Fig. 1.13).

In the figure, plasma total protein values are marked in grams per dL on the horizontal axis and plasma density is marked in grams per cm^3 on the vertical axis. The measurement points are aligned on a straight line. This shows us that there is a complete linear relationship between plasma total protein concentration and plasma density. The relationship can be expressed as $y = mx + n$.

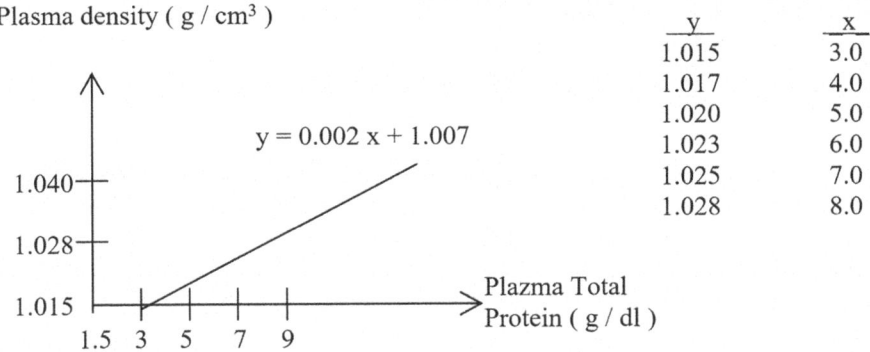

y	x
1.015	3.0
1.017	4.0
1.020	5.0
1.023	6.0
1.025	7.0
1.028	8.0

Fig. 1.13   Graph between plasma density and plasma total protein amount

The slope ratio calculated from the values ($\Delta x = 1$; $\Delta y = 0.0027$) found by measuring on the graph is

$$\Delta y / \Delta x = 0.002$$

Since the ratio of the change amounts in the case of linear change is equal to the slope parameter m = 0.002 and accordingly the relation is

$$y = 0.002\ x + n$$

As a known point, (x_1, y_1) can be written in the equation and n can be calculated. For example, if the pair (6.0; 1.023) is used,
1.023 = 0.002 × 6.0 + n ve n = 1.007 is found. In that case, the density, total protein relationship equation can be written as

$$y = 0.002\ x + 1.007$$

The fact that the obtained line does not pass through the beginning of the coordinate axes indicates that the change relationship does not show a proportional increase.

In addition to line graphs, equations expressing this relationship can also be found:

2. Equation of a Line Whose Slope and One Point It Passes Through Are Known

The equation of the line whose slope is known as m and whose point is **A** (x_1, y_1) can be expressed as

$$y - y_1 = m\ (x - x_1)$$

3. Equation of a Line That Has an Unknown Slope but Has Known Two Points That It Passes Through

The equation of the line passing through two points A (x_1, y_1) ve B (x_2, y_2) can be expressed in two ways:

1. First, the slope is found with the help of $\Delta y/\Delta x$ and then one of the points is taken, and when the slope is known, it is written in its place in the equation of the line used. It can be used at both points because if a point is on the curve, it satisfies the equation.
2. It can be found by substituting into the equation $y - y_1/y_1 - y_2 = x - x_1/x_1 - x_2$

4. Positions of Truth Relative to Each Other

Let us consider the equations $ax + by + c = 0$ and $áx + èy + ñ = 0$

1. Parallel Lines

Lines with equal slopes are parallel to each other. In other words, $m_1 = m_2$. From another perspective, when $a/á = b/è \neq c/ñ$, the lines are parallel.

2. Vertical Lines

Lines whose slopes are multiplied by -1 are perpendicular to each other. In other words, $m_1 . m_2$ must be -1.

3. Nonparallel Lines

The slopes of these types of lines are also equal. Unlike parallel lines, $a/á = b/è = c/ñ$

4. Secant Lines

Perpendicular lines are a special case of intersecting lines. The coordinates of the intersecting points can be easily found by solving the common equations of the given lines.

5. Finding the Middle Point

To find the midpoint of two given points,

$$x_0 = (x_1 + x_2)/2, \quad y_0 = (y_1 + y_2)/2 \quad \text{formulas are used.}$$

If it is a triangular area, these parameters are applied as

$$x_0 = (x_1 + x_2 + x_3)/3, \qquad y_0 = (y_1 + y_2 + y_3)/3$$

If it is desired to determine the area of this triangular region, the determinant rule can be used as an easy method. However, it should not be forgotten that the result obtained is equal to twice the area to be calculated.

6. Constant Function

It is the graph in which the y value does not change at all and remains the same according to the changing x values (Fig. 1.14).

Fig. 1.14 Constant graph

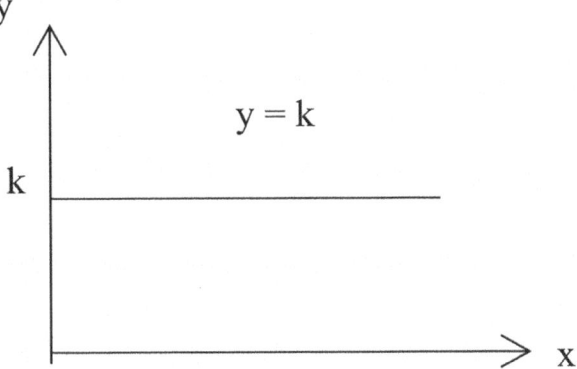

Fig. 1.15 Inverse proportion graph

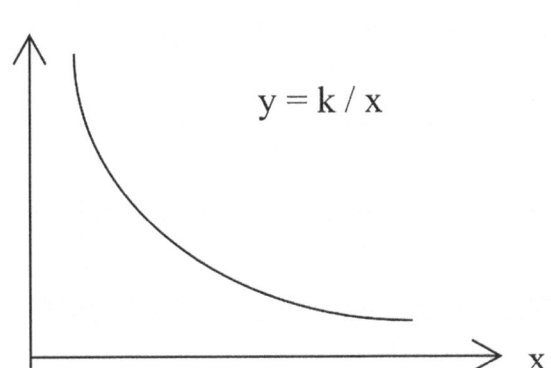

7. Inverse Proportion

If one of the quantities x and y increases while the other decreases, or if one decreases while the other increases, these two quantities are said to be inversely proportional and are denoted by y.x = k. Inversely proportional expressions are parameters in a multiplicative relationship. That is, if two quantities are inversely proportional, their product remains constant.

If the inversely proportional expressions are plotted according to certain values, this is a curvilinear graph (Fig. 1.15).

In mathematics, the graphs of functions defined by $y = ax^2 + bx + c$ are in the form of parabolas. Graphs in this form can be expressed in five different forms (Fig. 1.16):

1. $y = a x^2$

   It is the equation of the parabola passing through the starting point. If a > 0, the arms of the parabola open upward; if a < 0, the arms of the parabola open downward; and as |a| increases, the arms of the parabola narrow.

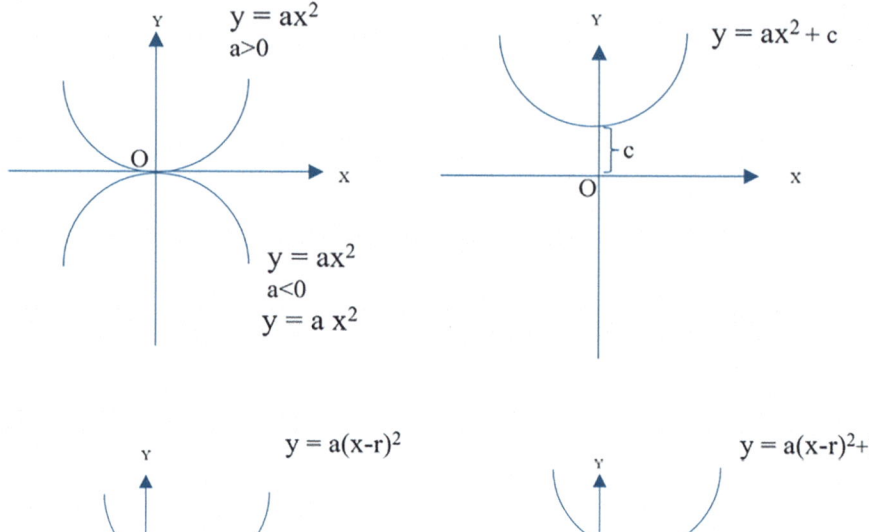

Fig. 1.16 Various parabola graphs

2. $y = a x^2 + c$

It is a parabola whose starting point is on the y-axis and whose outline can be drawn by moving it up or down from the origin by a distance of c. The parabola definitely intersects the y-axis at the point (0,c). In order to determine whether it intersects the x-axis, an examination should be made by giving $y = 0$.

3. $y = a (x - r)^2$

It is a parabola whose starting point is on the x-axis and whose outline can be drawn by moving it to the right or left from the origin by an amount r. These parabolas, which are tangent to the x-axis $(x - r = 0)$, are examined by giving $x = 0$ to see whether they intersect the y-axis.

4. $y = a (x - r)^2 + k$

They are parabolas with a vertex (r, k) and a starting point in the regions where r and k values are different from zero.

5. $y = ax^2 + bx + c$

In its most general sense, it is the expression of the equation of the parabola, and the coordinates of the vertex can be found with the help of the relevant formulas.

$$r = -b/2a, k = 4\,a\,c - b^2/4a$$

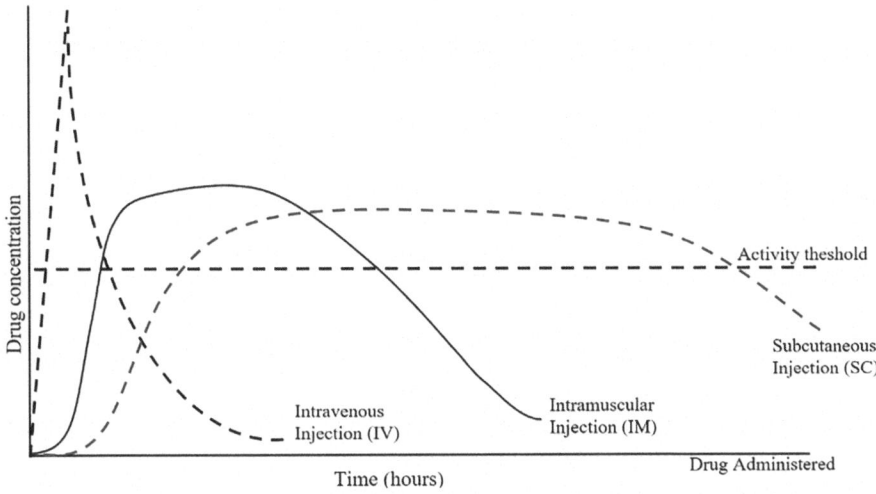

Fig. 1.17 Graph showing active ingredient concentrations with different injection routes

In addition, the expression given by the equation $(x - a)^2 + (y - b)^2 = r^2$ also determines a circle. The coordinates of the center point of this circle are defined as M (a, b) and its radius is defined as r.

For the equation $Ax^2 + B \times y + Cy^2 + D \times + E y + F = 0$ to represent a circle

1. B must $= 0$. That is, there should not be a term in the form of xy in the equation.
2. A must $= C$. That is, the coefficients of the terms containing x^2 and y^2 must be equal.
3. θ must $= a^2 + b^2 - F > 0$.

Apart from these, there may be various changes that do not comply with certain rules. For example, the relationship between the active ingredient concentration and the applied injection routes is examined in the graph below (Fig. 1.17).

Questions 5

1. The plasma density (g/cm^3) and plasma total protein (g/dL) values of a dog are given below. Find the equation expressing this change and draw the graph.

x	y
4	1.0169
5	1.0199
6	1.0229
7	1.0259
8	1.0289
9	1.0319

On the horizontal axis, plasma total protein values are plotted in grams per dL, and on the vertical axis, plasma density is plotted in g/cm^3. The relationship can be expressed as y = mx + n.

From the table, Δx = 1; Δy = 0.003 is found. Therefore

$$\Delta y / \Delta x = 0.003$$

Since the ratio of the change amounts in the linear change case is equal to the slope parameter, m = 0.003, and accordingly the relation is

$$y = 0.003\ x + n$$

n can be calculated by substituting (x_1, y_1) into the equation as a known point. For example, if the pair (4; 1.0169) is used

1.0169 = 0.003 × 4 + n and n = 1.0049 is found. Then the density, total protein relationship equation can be written as

$$y = 0.003\ x + 1.0049$$

Chart,

2. The plasma density (g/cm^3) and plasma total protein (g/dL) values of a dog are given below. Find the equation expressing this change and draw the graph.

x	y
2	1.0116
2.5	1.0192
3	1.0268
3.5	1.0344
4	1.0420
4.5	1.0496

On the horizontal axis, plasma total protein values are plotted in grams per dL, and on the vertical axis, plasma density is plotted in g/cm^3. The relationship can be expressed as y = mx + n.

From the table it is found that Δx = ... ; Δy = ... Therefore

$$\Delta y / \Delta x = ...$$

Since the ratio of the change amounts in the case of linear change is equal to the slope parameter, m = ... and accordingly the relation is

$$y = ...x + n$$

n can be calculated by substituting (x1, y1) into the equation as a known point. For example, if the pair (.........; ....) is used n = ... is found. Then

the density, total protein relationship equation can be written as y = ... x
+............ ..
   Chart,

3. What is the graph between hematocrit value and erythrocyte count?
4. What is the graph between the hematocrit value and the reticulocyte ratio, keeping the reticulocyte count constant?
5. I. $2x + 3y - 5 = 0$ II. $2x + 3y - 10 = 0$ III. $4x + 6y - 10 = 0$ IV. $4x + 6y - 5 = 0$

Which two of the lines whose equations are given above are coincident? Prove the result you found with the formulas by drawing the graph of each line.

One of the important places where we will use the concept of ratio and proportion in our medicine is the evaluation of the type of anemia, called the *binary nomenclature method*, which can be applied to various animal species. Total erythrocyte count (RBC), hematocrit value (PCV), and hemoglobin (HGB) values are examined to determine whether the patient has anemia. For this process, the normal limits of each value must be known. If these values have decreased, it can be called *anemia*; if they have increased, it can be called *polycythemia*. Total leukocyte count (WBC) and formula leukocyte values are parameters that allow us to comment on diseases. Platelet count (PLT) is a parameter that gives an idea about blood clotting.

To make these measurements, blood is definitely taken using veins. Arteries are not used for taking blood except for blood gas analysis. In horses, the V. jugularis, which runs in the neck, is very suitable for taking blood. However, it should be done very carefully because of the high risk of thrombophlebitis in horses. The same vein can be used in cows, as well as the V. coccygea in the tail and the V. subcutanea abdominalis (V. thoracica externa) under the abdomen. Only the V. jugularis can be used in sheep and goats. In dogs, blood can be easily taken from the V. cephalica antebrachii running on the front leg and the V. cephana parva (magnum) in the hind leg, while the V. jugularis is used if a large amount of blood is to be taken or if it is one of the small breeds. In cats, the V. jugularis is prepared for the blood collection process after being brought under control.

In addition, in allergic conditions such as urticaria and eczema encountered in dogs, cats, horses, and cows, *autohemotherapy* applications are also performed using these veins. For this purpose, approximately 10 cc of blood is taken from the dog and given to it subcutaneously. This is a proteinotherapy and the aim is to activate the immune system.

The same veins are also used for intravenous injections. Subcutaneous (s.c.) injections are generally made by making a fold in the skin over the scapula, thinking of an imaginary cone, and are made toward the very center of this. If necessary, the cannula can be rotated without removing the needle to ensure the spread of the fluid. It can also be applied to the neck area in cattle. However, in small ruminants such as sheep and goats, the armpit is preferred because the body is covered with wool. Intramuscular (I.M.) applications are made to the neck in large animals and to the space between the semitendinosus and semimembranosus in carnivores. In recent years, due to the risk of phlebitis, even I.M. injections in horses are made to the chest muscles instead of the neck. The skin of the same area is also used for the s.c. route.

Blood is used either with anticoagulant (containing a substance that prevents clotting) or without anticoagulant. If some anticoagulant substances such as EDTA, oxalate, and heparin are mixed into the blood, this blood does not clot and is used to determine hemogram parameters. If blood is taken without anticoagulant, serum is obtained and the serum is used for various biochemical and microbiological tests. For this purpose, the blood is taken directly into a tube, kept without shaking, and if necessary, centrifuged and waited for the serum to accumulate at the top. Serum is a light-colored, clear liquid. If there is redness, hemolysis has occurred and this reduces the reliability of the measurement results. If it is yellow, icterus (jaundice) is suggested. However, if the patient is a horse, the normal serum color is already yellow and is not related to icterus.

8. Normal PCV Values in Adult Animals (%)

Dog	37–55
Cat	24–45
Horse	32–52
Cow	26–42
Sheep	24–29

9. Normal HGB Values in Adult Animals (g/100 mL)

Dog	11.7–14.9
Cat	8.1–13.5
Horse	8.5–13
Cow	8.5–13.5
Sheep	8.3–14.8

Researchers report that as a rule of thumb, the hemoglobin concentration is roughly 1/3 of the hematocrit value.

10. Normal RBC Values in Adult Animals ($\times 10^6/mm^3$)

Dog	5.5–8.5
Cat	5.0–10.0
Horse	6.0–12.0
Cow	5.0–10.0
Sheep	6.5–11.3

When using each of these values, we must always consider the lower and upper limits and never use the average value. If none of these values are below the normal level, that is, if the case is not anemic, there is no point in applying the binomial nomenclature method.

11. Binary Nomenclature Method

In this method, the first nomenclature is made using the mean erythrocyte volume (MCV).

$$MCV \text{ (fl)} = \frac{PCV \times 10}{RBC} \qquad \text{formula is used.}$$

If the MCV value is within the normal limits of the given animal species, it is called normocytic anemia; if it is above it, it is called macrocytic anemia; and if it is below it, it is called microcytic anemia.

In this method, the second nomenclature is made using the mean corpuscular hemoglobin (MCH) or the mean corpuscular hemoglobin volume (MCHC).

$$MCH \text{ (pg)} = \frac{HGB \times 10}{RBC} \qquad MCHC \text{ (g/dL)} = \frac{HGB \times 100}{PCV} \qquad \text{formula is used.}$$

If the MCH or MCHC value is within the normal limits of the given animal species, it is called normochromic anemia; if it is above it, it is called hyperchromic anemia; and if it is below it, it is called hypochromic anemia.

MCV, MCH, and MCHC values for various animal species are given below:

12. Normal MCV Values in Adult Animals (fl)

Dog	60–77
Cat	40–55
Horse	37–55
Cow	46–65
Sheep	34–46

13. Normal MCH Values in Adult Animals (pg)

Dog	17–23
Cat	13–17
Horse	13–19
Cow	11–17
Sheep	13–14

14. Normal MCHC Values in Adult Animals (g/dL)

Dog	31–34
Cat	31–35
Horse	31–36
Cow	31–34
Sheep	29–34

When all of this information is evaluated, anemias can be classified as follows:

A. Normocytic-Normochromic Anemia

In this type of anemia, the erythrocytes are of normal size and the hemoglobin they contain is at normal levels. It most commonly occurs as a result of acute bleeding.

B. Microcytic-Hypochromic or Normochromic Anemia

In this type of anemia, the erythrocytes are smaller than normal and the amount of hemoglobin they contain is lower than normal. It most commonly occurs as a result of Fe deficiency (dietary, chronic bleeding, pregnancy), Cu deficiency, and Mo poisoning. In cat and dog practice, it can also occur in some breed and age-related factors.

C. Macrocytic-Hyperchromic or Normochromic Anemia

In this type of anemia, the erythrocytes are larger than normal and the amount of hemoglobin they contain is higher than normal. It most commonly occurs as a result of vitamin B12 deficiency (pernicious anemia), Co deficiency in ruminants, and leukemia virus (FeLV) infections in cats.

In addition, while there was no anemic picture, small erythrocytes (MCV = 55–65) were observed in some dog breeds (Japanese Akita) and larger than normal erythrocytes (MCV = greater than 80 fl) were observed in some (poodle).

It has been found that MCV values are high in newborn cats and dogs but decrease to normal levels within 2–3 months in dogs and 2 months in cats.

D. Color Index

Another method to determine the type of anemia is the calculation of the color index. For this purpose, first,

$$\frac{\text{Measured HGB}}{\text{Normal HGB}} \text{ is calculated. Then, } \frac{\text{Measured RBC}}{\text{Normal RBC}} \text{ rate is found.}$$

It is determined as $C.I = \frac{\text{Hemoglobin related ratio}}{\text{RBC related ratio}}$. The interpretation of this value is as follows:

If C.I. > 1.0, it is hyperchromic anemia. It is observed in hemolytic anemias.
If C.I. = 1.0, it is normochromic anemia.
If C.I < 1.0, it is hypochromic anemia and is observed in nutritional anemias.

E. RDW and Reticulocytes

The parameter specified as RDW is the percentage expression of the volumetric difference of the erythrocytes. If RDW decreases, it is understood that the population is dominated by erythrocytes with similar diameters. If RDW increases, it is thought that there are erythrocytes with large diameter differences. This situation is mostly observed in relation to nutritional deficiencies and inflammation. The RDW parameter should be evaluated together with MCV. If RDW increases together with MCV, this situation suggests regenerative anemias, for example, hemorrhagic anemias.

Reticulocytes are known as young erythrocytes and their increase is called reticulocytosis. This condition is also observed in regenerative anemias. In other words, the reticulocytosis picture occurs in regenerative anemias. Non-regenerative anemias are more related to the decrease or absence of bone marrow RBC capacity. Secondary-type anemias are also known as non-regenerative-type anemias.

F. Interpretation of Total Leukocyte and Formula Leukocyte Values

An increase in the total leukocyte value is called leukocytosis, while a decrease is called leukopenia. Leukocytosis is more commonly observed in inflammation, stress, exercise, and tumoral events where blood cells in the bone marrow and other tissues increase excessively. Leukopenia is seen in inflammatory diseases and bone marrow diseases where there is excessive tissue consumption. A decrease in the number of all types of leukocytes is also defined as panleukopenia.

1. Normal WBC Values in Adult Animals ($\times 10^3/\mu L$)

Dog	6–12
Cat	6–11
Horse	5–10
Cow	5–10
Sheep	4–6

Leukocytes are divided into two main groups: polymorphic nuclear leukocytes (granulocytes) and mononuclear leukocytes (agranulocytes).

(a) Granulocytes: They are multinucleated leukocytes. Neutrophils, eosinophils, and basophils are included in this group.
(b) Agranulocytes: They are leukocytes with a single nucleus. Lymphocytes and monocytes are included in this group.

In cases of leukocytosis or leukopenia, the leukocyte formula is prepared to determine which cell or cells have changed in number. For this purpose, a blood smear is prepared.

When evaluated in terms of leukocytosis, the increase in neutrophil leukocytes is called neutrophilia, the increase in eosinophil leukocytes is called eosinophilia, the increase in basophils is called basophilia, the increase in lymphocytes is called lymphocytosis, and the increase in monocytes is called monocytosis.

When evaluated in terms of leukopenia, the decrease in neutrophil leukocytes is called neutropenia, the decrease in eosinophil leukocytes is called eosinopenia, the decrease in basophils is called basofilopenia, the decrease in lymphocytes is called lymphopenia, and the decrease in monocytes is called monocytopenia.

Neutrophilia mainly occurs as a result of bacterial infections, fear, cold, and excessive muscle work; neutropenia occurs in viral infections.

While eosinophilia mainly occurs in parasitic invasions and allergic diseases, eosinopenia usually occurs as a result of stress and steroid effects.

Basophilia mainly occurs as a result of hypersensitivity reactions, tumors, and parasites, especially blood parasites, while basofilopenia occurs as a result of hereditary causes and hyperthyroidism and anaphylactic conditions.

Lymphocytosis occurs as a result of some parasites settling in the lymph nodes and lymphosarcoma; lymphopenia occurs in acute systemic infections, chronic renal failure, and stress.

While monocytosis is observed in the initial stages of acute infections, monocytopenia occurs in viral infections, especially in large animals.

2. Normal Formula Leukocyte Values in Adult Animals

	Horse	Cow	Sheep	Dog	Cat
Neutrophil	50–64	15–45	15–34	19–81	35–75
Lymphocyte	20–40	40–70	35–75	15–30	20–56
Basophil	0–2	0–1	0–1	0–1	0–1
Eosinophil	4–10	2–20	0–10	2–8	1–12
Monocyte	3–10	3–10	0–6	1–8	0–4

When evaluating the patient, 100 cells are counted and the ratio is made accordingly. There is also another evaluation table called the leukosis key, which is used only in ruminants for leukosis. With the help of this key, it can be said whether the animal has leukosis or is suspicious. The criteria to be evaluated are the age of the animal, total leukocyte value, and lymphocyte ratio.

If the cow is 2 years old or younger,

Leukosis	Total leukocyte ($\times 10^3/\mu L$)	Lymphocyte ratio (%)
Negative	<12	<60
Suspicious	12–18	60–75
Positive	18<	75<

If the cow is more than 2 years old,

Leukosis	Total leukocyte ($\times 10^3/\mu L$)	Lymphocyte ratio (%)
Negative	<10	<65
Suspicious	10–18	65–75
Positive	18<	75<

Questions 6

1. RBC $= 4.2$ $(\times 10^6/mm^3)$
    PCV $= 13$ $(\%)$
    HGB $= 3.3$ (g/dL)
    WBC $= 9000$ $(/\mu L)$

The blood laboratory findings of a randomly selected sheep from the flock in which a parasite called Fasciola hepatica was detected as a result of the autopsy are given above. Answer the following questions accordingly.

(a) Prove that it has anemia by writing down the normal values of RBC, PCV, and HGB.

Since our patient is a sheep, let's write down the normal RBC, PCV, and HGB values for a sheep.

$$RBC = 6.5 - 11.3 \left(\times 10^6/mm^3 \right)$$
$$PCV = 24 - 29 \quad (\%)$$
$$HGB = 8.3 - 14.8 \ (g/dL)$$

As can be seen, the RBC, PCV, and HGB values of the sick sheep are low compared to normal values. Therefore, our case is anemic.

(b) Find the type of anemia using the binomial nomenclature method by writing the necessary formulas.

$$\text{MCV (fl)} = \frac{\text{PCV} \times 10}{\text{RBC}} = \frac{13 \times 10}{4.2} = \frac{130}{4.2} = 30.95$$

The value found is low because the normal MCV values for sheep are between 34 and 46. Therefore, our case has a microcytic anemia.

$$\text{MCH (pg)} = \frac{\text{HGB} \times 10}{\text{RBC}} = \frac{3.3 \times 10}{4.2} = \frac{33}{4.2} = 7.85$$

The value found is low because the normal MCH values for sheep are between 13 and 14. Therefore, our case has a hypochromic type of anemia.

If we combine these findings according to the binomial nomenclature method, we reach the conclusion that the sheep has a microcytic, hypochromic type of anemia.

(c) Write the normal value of WBC and write what the change is called.

The WBC value for sheep is normally between 4 and 6 ($\times 10^3/\mu$L). Since the WBC value in our patient was 9 ($\times 10^3/\mu$L), this change is called leukocytosis.

(d) Calculate and interpret the color index value.

For this purpose, first find the $\dfrac{\text{Measured HGB}}{\text{Normal HGB}} = \dfrac{3.3}{11.55} = 0.28$ value.

The normal HGB value was found to be 8.3 + 14.8 = 23.1/2 = 11.55 g/dL.

Then the $\dfrac{\text{Measured RBC}}{\text{Normal RBC}} = \dfrac{4.2}{8.9} = 0.47$ value is found.

The normal RBC value was found to be 6.5 + 11.3 = 17.8/2 = 8.9 ($\times 10^6$/ mm^3).

$$\text{C.I} = \frac{\text{Hemoglobin related ratio}}{\text{RBC related ratio}} = \frac{0.28}{0.47} = 0.59$$

When we compare the color index value we found with 1.0, we see that our value is less than 1. Therefore, we can talk about a nutritional anemia.

(e) What kind of change is observed in which cell type in the leukocyte formula? Name this change.

Since a parasite called Fasciola hepatica is detected as a result of the autopsy, we observe an increase in the number of eosinophils, especially in the formula leukocyte. This is called eosinophilia.

(f) From which vein could blood have been taken? Also write the location.

Since our patient is a sheep, the jugular vein can be used. This vein runs in the neck.

(g) Where is the subcutaneous injection given to this animal?

S.c. injection can be made between the front and back legs of sheep since their entire body is covered with wool.

2. RBC = 2.5 ($\times 10^6$ mm^3)   PCV = 24 (%)   HGB = 8.5 (g/dL)

Blood analysis of a 5-year-old female dog is given above.

A. What type of blood should be taken for these measurements?

Anticoagulant blood should be taken for RBC, PCV, and HGB values.

B. From which veins could the blood have been taken? (Give three different places and their names.)

Vena cephalica antebrachium   (front leg)
Vena cephana parva   (hind leg)
Vena jugularis   (neck)

C. If it has anemia, explain its type and how you found it.

Since our patient is a dog, let's write down the normal RBC, PCV, and HGB values for the dog.

$$RBC = 5.5 - 8.5 \quad (\times\ 10^6/mm^3)$$
$$PCV = 37 - 55 \quad (\%)$$
$$HGB = 11.7 - 14.9 \quad (g/dL)$$

As can be seen, the RBC, PCV, and HGB values of the sick dog are low compared to normal values. Therefore, our case is anemic.

With the binomial nomenclature method, we find the type of anemia by writing the necessary formulas.

$$MCV\ (fl) = \frac{PCV \times 10}{RBC} = \frac{24 \times 10}{2.5} = \frac{240}{2.5} = 96$$

The value found is high because the normal MCV values for dogs are between 60 and 77. Therefore, our case has a macrocytic anemia.

$$\text{MCH (pg)} = \frac{\text{HGB} \times 10}{\text{RBC}} = \frac{8.5 \times 10}{2.5} = \frac{85}{2.5} = 34$$

The value found is high because the normal MCH values for dogs are between 17 and 23. Therefore, our case has a hyperchromic type of anemia.

If we combine these findings according to the binomial nomenclature method, we conclude that the dog has a macrocytic, hyperchromic type of anemia.

D. What diseases do you suspect?

Vitamin B_{12} deficiency may be considered.

3. RBC = 3.5 (x10^6 mm^3) PCV = 24 (%)   HGB = 8.5 (g/dL)

Blood analysis of a 3-year-old female cat is given above.

A. Does the cat have anemia? How did you understand?

Since our patient is a cat, let's write down the normal RBC, PCV, and HGB values for the cat.

RBC = ...$\left(\times 10^6 / \text{mm}^3 \right)$
PCV = ...(%)
HGB = ...(g/dL)

As can be seen, the RBC, PCV, and HGB values of the sick cat are low compared to normal values. Therefore, our case is anemic.

B. Explain the type and how you found it.

With the binomial nomenclature method, we find the type of anemia by writing the necessary formulas.

MCV (fl) =

Since the normal MCV values for cats are between ...., the value found is...................... Therefore, our case has a .... type of anemia.

MCH (pg) =

Since the normal MCH values for the cat are between ...., the value found is .... Therefore, our case has a .... type of anemia.

If we combine these findings according to the binomial nomenclature method, we reach the conclusion that the cat has an ...., ... type of anemia.

C. What diseases do you suspect?

....

4. RBC $= 3.89$ ($\times 10^6$ mm^3)   PCV $= 26(\%)$ HGB $= 8.3$ (g/dL)

Blood analysis of a 3-year-old female dog is given above.

(a) What type of blood should be taken for these measurements?

....

(b) From which veins could the blood have been taken? (Give three different places and their names.)

.... (neck)

.... (front leg)

.... (hind leg)

(c) If she has anemia, explain the type and how you found it.

Since our patient is a dog, let's write down the normal RBC, PCV, and HGB values for the dog.

$$RBC = ... \left(\times 10^6 / mm^3 \right)$$
$$PCV = ... (\%)$$
$$HGB = ... (g/dL)$$

As can be seen, the RBC, PCV, and HGB values of the sick dog are low compared to normal values. Therefore, our case is anemic.

With the binomial nomenclature method, we find the type of anemia by writing the necessary formulas.

$$MCV \ (fl) =$$

Since the normal MCV values for the dog are between ...., the value found is....................... Therefore, our case has a .... type of anemia.

$$MCH \ (pg) =$$

Since the normal MCH values for the dog are between .... the value found is .... Therefore, our case has a .... type of anemia.

If we combine these findings according to the binary nomenclature method, we reach the conclusion that the dog has an ...., ... type of anemia.

5. The blood analysis results of a 5-year-old male horse are given below. Answer the questions by making calculations based on this data.

$$\begin{aligned}
&\text{RBC} && 5.5 \ (\times 10^6/\text{mm}^3) \\
&\text{PCV} && 17.0 \ (\%) \\
&\text{WBC} && 13.5 \ (10^3/\text{mm}^3) \\
&\text{HGB} && 2.8 \ (\text{g/dL})
\end{aligned}$$

		Normal values
Neutrophil leukocyte	60	(35–75)
Eosinophil leukocyte	16	(1–12)
Lymphocyte	20	(20–56)
Basophil	0	(0–1)
Monocyte	4	(0–4)

(a) What type of blood should be taken for these measurements?
   Anticoagulant blood should be taken.
(b) Write the names of the veins used for this process, indicating their locations.
   Vena jugularis
(c) How were the formula leukocyte values obtained by preparing the blood?
   A blood smear is prepared.
(d) Prove that it has anemia by writing the normal values.

Since our patient is a horse, let's write down the normal RBC, PCV, and HGB values for a horse.

$$\begin{aligned}
&\text{RBC} = 6.0 - 12.0 && \left(\times 10^6/\text{mm}^3 \right) \\
&\text{PCV} = 32 - 52 && (\%) \\
&\text{HGB} = 8.5 - 13 && (\text{g/dL})
\end{aligned}$$

As can be seen, the RBC, PCV, and HGB values of the sick horse are low compared to normal values. Therefore, our case is anemic.

(e) Find the type of anemia by making the necessary calculations.

$$\text{MCV (fl)} = \frac{\text{PCV} \times 10}{\text{RBC}} = \frac{17 \times 10}{5.5} = \frac{170}{5.5} = 30.90$$

The value found is low because the normal MCV values for the horse are between 37 and 55. Therefore, our case has a microcytic anemia.

$$\text{MCH (pg)} = \frac{\text{HGB} \times 10}{\text{RBC}} = \frac{2.8 \times 10}{5.5} = \frac{28}{5.5} = 5.09$$

Since the normal MCH values for the horse are between 13 and 14, the value found is low. Therefore, our case has a hypochromic type of anemia.

If we combine these findings according to the binomial nomenclature method, we reach the conclusion that the horse has a microcytic, hypochromic type of anemia.

(f) What can be considered as the cause of anemia for the case? (Write only one)
   Fe deficiency
(g) Name the change in total leukocyte value.

The WBC value for a horse is normally between 5 and 10 ($\times 10^3/\text{mm}^3$). Since the WBC value in our patient was 13.5 ($\times 10^3/\text{mm}^3$), this change is called leukocytosis.

(h) Name the change in the formula leukocyte value.

When the table is examined, we see that the number of eosinophils, especially the leukocyte formula, increases. This is called eosinophilia.

(i) Write only one name for the change in the leukocyte type in the formula for the case.
   Parasitic diseases

6. The blood analysis results of a 5-year-old female cat are given below. Answer the following questions by making calculations based on this data.

RBC	3.5	($\times 10^6/\text{mm}^3$)
PCV	22.0	(%)
WBC	4.5	($\times 10^3/\text{mm}^3$)
HGB	7.9	(g/dL)

	Normal values	
Neutrophil leukocyte	30	(35–75)
Eosinophil leukocyte	12	(1–12)
Lymphocyte	54	(20–56)
Basophil	0	(0–1)
Monocyte	4	(0–4)

(a) What type of blood should be taken for these measurements?

....

(b) Write the names of the veins used for this process, indicating their locations.

.... (...........................)

.... (...........................)

(c) How were the formula leukocyte values obtained by preparing the blood?

....

(d) Prove that he/she has anemia by writing down the normal values.

Since our patient is a cat, let's write down the normal RBC, PCV, and HGB values for the cat.

$$RBC = ... \left(\times 10^6 / mm^3 \right)$$
$$PCV = ... (\%)$$
$$HGB = ... (g/dL)$$

As can be seen, the RBC, PCV, and HGB values of the sick cat are low compared to normal values. Therefore, our case is anemic.

(e) Find the type of anemia by making the necessary calculations.

$$MCV \ (fl) =$$

Since the normal MCV values for cats are between ...., the value found is...................... Therefore, our case has a .... type of anemia.

$$MCH \ (pg) =$$

Since the normal MCH values for cats are between ...., the value found is .... Therefore, our case has a .... type of anemia.

If we combine these findings according to the binomial nomenclature method, we reach the conclusion that the cat has an ...., ... type of anemia.

(f) What can be considered as the cause of anemia for the case? (Write only one)

....

(g) Name the change in total leukocyte value.

The WBC value for cats is normally between .... $(\times 10^3 / mm^3)$.

Since the WBC value in our patient was...................... $(\times 10^3 / mm^3)$, this change is called ....

(h) Name the change in the formula leukocyte value.

....

(i) Write a disease name for the case where we can see the change in the leukocyte-type formula.

....

(j) Write down the normal pulse value for the case.

..................................(/minute)

7. The blood analysis results of a 5-year-old cow are given below. Answer the questions by making calculations based on this data.

RBC	3.5	$(\times 10^6/mm^3)$
PCV	17.0	(%)
WBC	16.0	$(\times 10^3/mm^3)$
HGB	5.5	(g/dL)

	Normal values	
Neutrophil leukocyte	20	(35–75)
Eosinophil leukocyte	2	(1–12)
Lymphocyte	75	(20–56)
Basophil	0	(0–1)
Monocyte	3	(0–4)

(a) Prove that he/she has anemia by writing down the normal values.

Since our patient is a cow, let's write down the normal RBC, PCV, and HGB values for a cow.

RBC = ...$\left(\times 10^6/mm^3 \right)$

PCV = ...(%)

HGB = ...(g/dL)

As can be seen, the RBC, PCV, and HGB values of the sick cow are low compared to normal values. Therefore, our case is anemic.

(b) Find the type of anemia by making the necessary calculations.

MCV (fl) =

Since the normal MCV values for the cow are between .... the value found is .... Therefore, our case has a .... type of anemia.

MCH (pg) =

Since the normal MCH values for the cow are between ...., the value found is .... Therefore, our case has a .... type of anemia.

If we combine these findings according to the binomial nomenclature method, we reach the conclusion that the cow has an ...., ... type of anemia.

(c) Name the change in total leukocyte value.
    The WBC value for a cow is normally between .... ($\times 10^3/mm^3$).
    Since the WBC value in our patient is .... ($\times 10^3/mm^3$), this change is called ....
(d) Name the changes in the formula leukocyte value.
    ....
    ....

(e) Comment on the condition of this patient in terms of leukosis using the leukosis key.

Since our patient is 5 years old, we should use the table valid for people over 2 years old. Our case is evaluated as suspicious because the total leukocyte count is 16.0 ($\times 10^3/mm^3$), between 12 and 18 ($\times 10^3/mm^3$), and the lymphocyte ratio is between 75 (%), 65–75 (%).

(f) Write the color index formula and comment on the type of anemia.

$$\text{For this purpose, first the} \quad \frac{\text{Measured HGB}}{\text{Normal HGB}}$$

= .... value is found.

The normal HGB value was found to be .... (g/dL).

$$\text{Then find the} \quad \frac{\text{Measured RBC}}{\text{Normal RBC}} = ...\text{value.}$$

The normal RBC value was found to be .... ($\times 10^6$ / mm^3).

$$C.I = \frac{\text{Hemoglobin related ratio}}{\text{RBC related ratio}} =$$

When we compare the color index value we found with 1.0, we see that our value is .... Therefore, ... anemia can be mentioned.

(g) From which veins could blood have been taken? Also write their locations. (Three pieces)

.... (neck)

.... (tail)

.... (underbelly)

(h) Which device was used to tranquilize this animal?

....

8. In a cow

$$RBC = 3.2 \quad (\times 10^6/mm^3)$$
$$HGB = 7.0 \quad (g/dL)$$

Determine the type of anemia using the color index.

For this purpose, first the, $\dfrac{\text{Measured HGB}}{\text{Normal HGB}} = ...$ value is found.

The normal HGB value was found to be .... g/dL.

Then find the $\dfrac{\text{Measured RBC}}{\text{Normal RBC}} = ...$ value.

The normal RBC value was found to be .... ($\times 10^6/mm^3$).

$$C.I = \frac{\text{Hemoglobin related ratio}}{\text{RBC related ratio}} =$$

When we compare the color index value we found with 1.0, we see that our value is .... Therefore, ... anemia can be mentioned.

9. Total leukocyte count of a 1-year-old cow was determined as 19,000 (/mm^3) and lymphocyte ratio was determined as 85 (%). Interpret the clinical condition using the leukosis key in terms of leukosis.

Since our patient is 1 year old, we should use the table valid for children under 2 years old. Since the total leukocyte count of our case is 19.0 ($\times 10^3/mm^3$), ... ($\times 10^3/mm^3$) is .... and the lymphocyte rate is .... from 85 (%), ... (%) it is evaluated as ....

10. A 3-year-old cow's total leukocyte count was determined as 10,000 (/mm^3) and the lymphocyte ratio was determined as 65 (%). Interpret the clinical condition using the leukosis key in terms of leukosis.

Since our patient is 3 years old, we should use the table valid for people over 2 years old. Since the total leukocyte count of our case is 10.0 ($\times 10^3/mm^3$), ... ($\times 10^3/mm^3$) is .... and the lymphocyte ratio is .... from 65 (%), ... (%), it is evaluated as ....

3. Concept of Function (Changing of Quantities Depending on Each Other)

A curve obtained by drawing the known path in relation to the measurement points, according to the perpendicular coordinate axes (Oy and Ox) marked on the analytical plane, determines a relation between two quantities. The value that the second quantity will take for a particular value of one of the quantities can be found with the help of this curve. Such a process is shown in Fig. 1.18.

The V1 value of the volume corresponding to the specific P1 value of the pressure is found with the help of perpendiculars. In mathematical terms, the P,V curve determines V as a function of P and is shown in mathematics as V=V(P) or V=f (P) and the quantity V is said to be a function of the variable P. In most cases, it is possible to find a mathematical equation that gives the curve. If the equation V=f (P) is found, the P_1 value can be calculated as $V_1 = f(P_1)$.

The general gas laws also change in a similar way (Figs. 1.19, 1.20, and 1.21).

$$P_1 V_1/T_1 = P_2 V_2/T_2$$

Fig. 1.18 Graph showing the change in pressure and volume of a gas at constant temperature

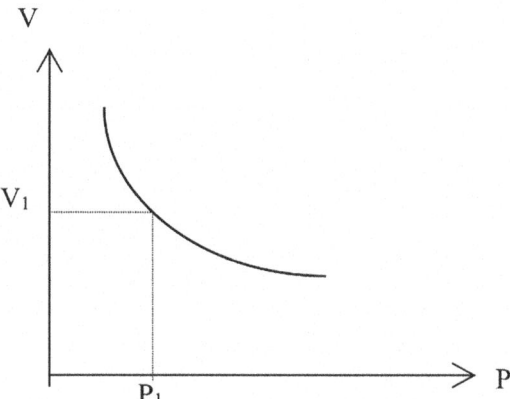

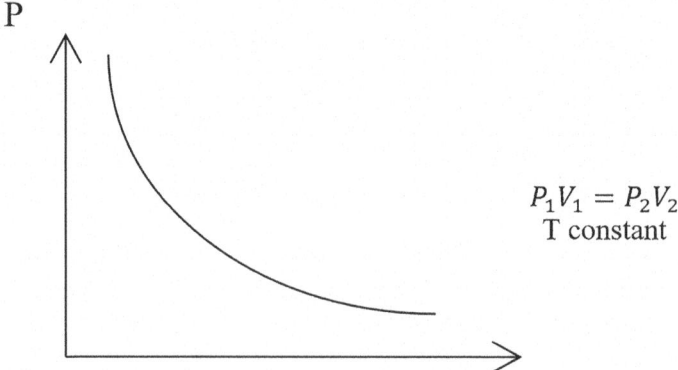

$$P_1 V_1 = P_2 V_2$$
$$T \text{ constant}$$

Fig. 1.19 Boyle-Mariotte law

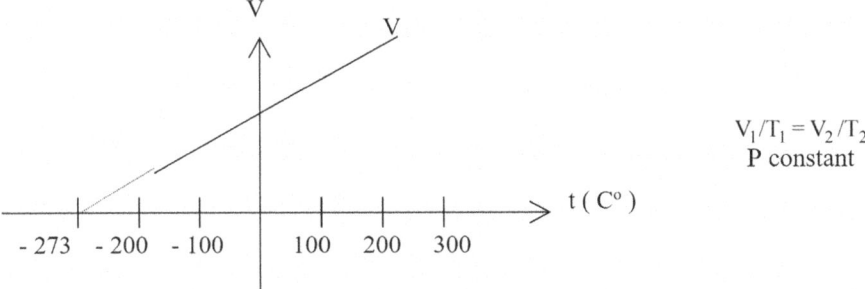

Fig. 1.20 Charles law

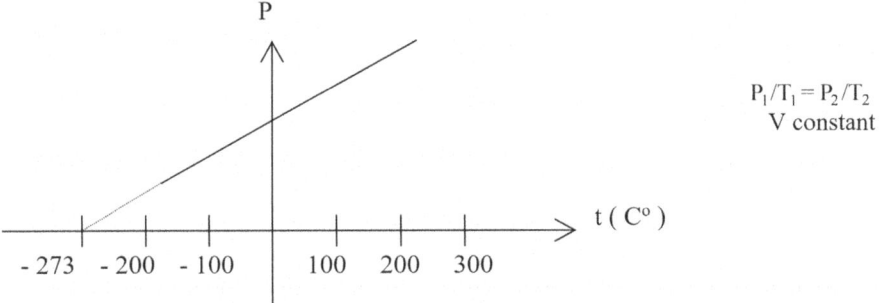

Fig. 1.21 Gay-Lussac law

Questions 7

1. Draw the graph between the pressure-volume product and the number of moles of the gas at a certain temperature.
2. Draw the graph between volume and number of moles while keeping pressure and temperature constant.
3. What is the graph of the product of pressure and volume vs. volume?
4. What is the pressure vs. density graph?
5. A certain mass of gas is heated while keeping its pressure constant. What is the relationship between the absolute temperature of the gas and its density?

G. Trigonometric Functions

The most commonly used trigonometric functions and what they represent are listed in Fig. 1.22.

The circle whose center is the starting point of the coordinate axes and whose radius is 1 unit is called the Trigonometry Circle. The axes divide this circle into four different regions. The signs of the trigonometric functions are indicated in Fig. 1.23.

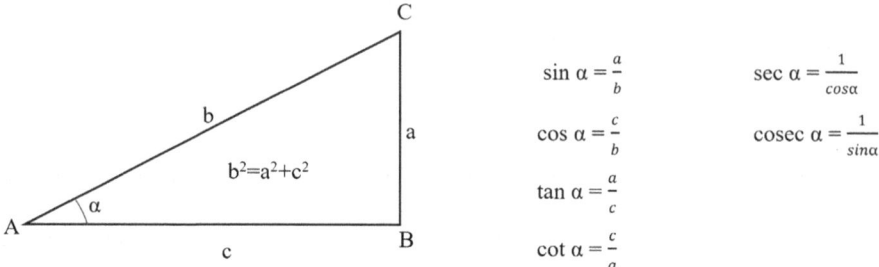

Fig. 1.22 Trigonometric triangle and trigonometric functions

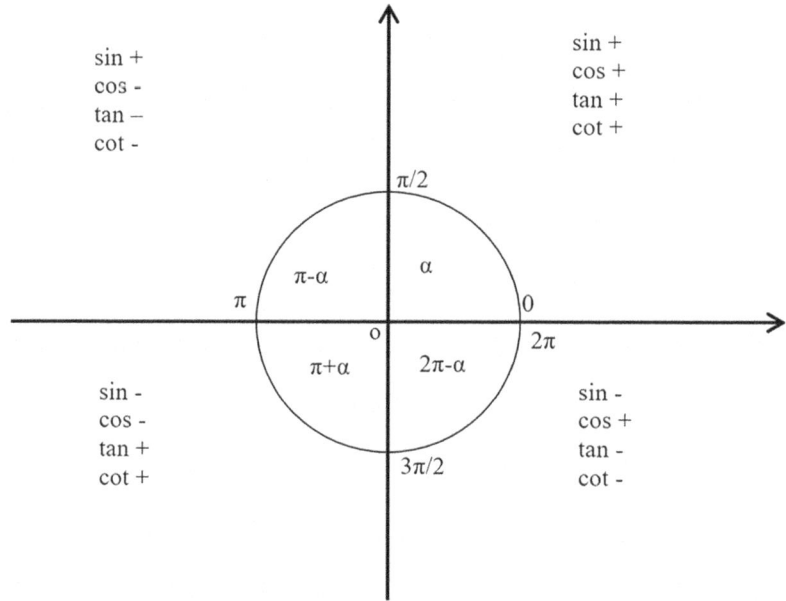

Fig. 1.23 Signs of trigonometric functions in the analytical plane according to regions and general rules for writing the angle in each region

Using these figures, the trigonometric values of other angles can be calculated with the help of the expressions whose numerical values are given in the table (Table 1.9).

Trigonometric functions are cyclic functions. Many periodic events are encountered in physics. The time change graph of some quantities gives a periodic curve. In some cases, this curve is only in the form of a sine or cosine curve. Sometimes, the shape of the periodic curve can be in any shape as in the figure. In physics, the meaning and reasons for this shape are always a subject of research. The mathematical analysis of periodic functions was developed by Fourier and is expressed by Fourier's theorem. Accordingly, a periodic function in any form of graph (Fig. 1.24)

Table 1.9 Values of trigonometric functions with respect to different limit values

	0	π/6	π/4	π/3	π/2	π	3π/2
x	0	30	45	60	90	180	270
Sin x	0	1/2	√2/2	√3/2	1	0	−1
Cos x	1	√3/2	√2/2	1/2	0	−1	0
Tan x	0	1/√3	1	√3	Undefined	0	Undefined
Cot x	Undefined	√3	1	1/√3	0	Undefined	0

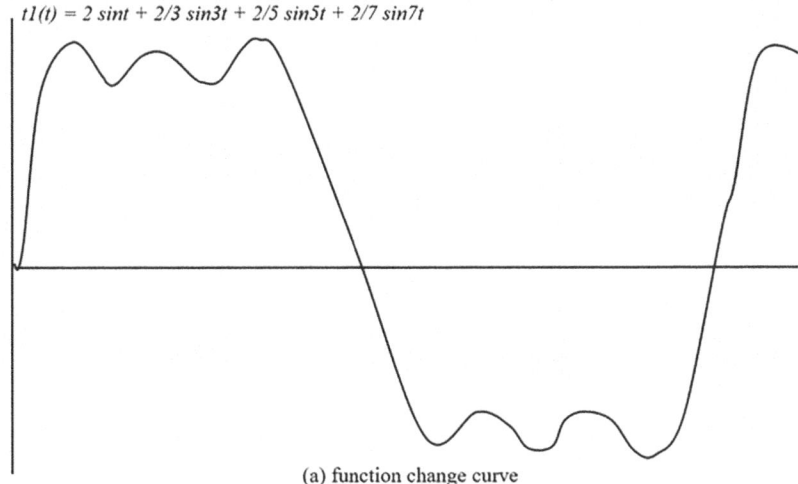

t1(t) = 2 sint + 2/3 sin3t + 2/5 sin5t + 2/7 sin7t

(a) function change curve

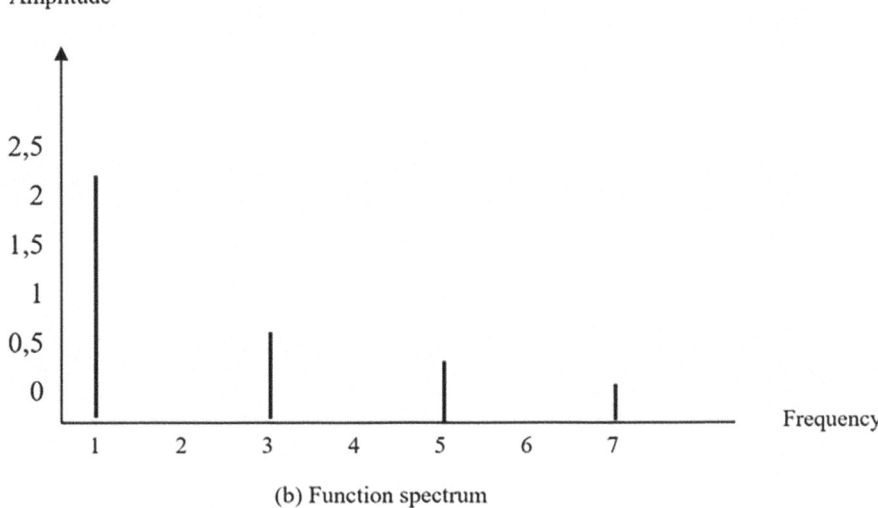

Amplitude

(b) Function spectrum

Fig. 1.24 Periodic function sample graph spectrum analysis

can be decomposed into many harmonic sine and cosine terms and can be expressed as the sum of these terms, this decomposition is possible only in one way.

A periodic function with period T is written as

$$f(t) = A_0 + A_1 \cos \frac{2\pi}{T}t + A_2 \cos 2\frac{2\pi}{T} + A_3 \cos 3\frac{2\pi}{T}t + \ldots$$

$$\ldots + B_1 \sin \frac{2\pi}{T}t + B_2 \sin 2\frac{2\pi}{T} + B_3 \sin 3\frac{2\pi}{T}t + \ldots$$

The constants A and B are the amplitudes given to the harmonic terms. Thanks to Fourier analysis, ECG and EEG observation graphs are evaluated as they are periodic functions and thus disease diagnosis methods are examined.

Questions 8

1. How many radians are 135°?
2. How many degrees are in $2\pi/3$ radians?
3. Calculate the trigonometric values given below, considering the regions and known angles.

(a) sin 330
    It is an angle in quadrant IV. It is written as $(2\pi - \alpha)$. Therefore,
    sin 330 = sin $(2\pi - 30)$
    In region IV, the sinus has a negative sign.
    sin $(2\pi - 30)$ = $-$ sin 30 = $-1/2$

(b) tan 225
    It is an angle in the third quadrant. It is written as $(\pi + \alpha)$. Then,
    tan 225 = tan $(\pi + 45)$
    In quadrant III, the tangent has a positive sign.
    tan $(\pi + 45)$ = tan 45 = 1

(c) cos 135

(d) cot 210

(e) sin 240

(f) cos 300

(g) tan 120

(h) cos $(3\pi/2 - \alpha)$

It is an angle in the third region. Here, the cosine is negative. So if the name change is made,

$$\cos(3\pi/2 - \alpha) = -\sin\alpha$$

(i) tan$(\pi/2 + \alpha)$

It is an angle in the second region. Here, the tangent is negative. So if the name change is made,

$$\tan (\pi/2 + \alpha) = - \cot \alpha$$

(j) sin (3 π/2 + α)=
(k) cot (π/2 − α)=
(l) tan (3π/2 − α)=
(m) cos (π/2 + α)=
(n) cos (−240)=

Angles with negative signs are affected by the sign in accordance with the rule of region IV. In this region, cosine is positive. Therefore,

$$\mathrm{Cos}\ (-240) = \cos 240$$

Now the angle is in quadrant III. Here, it is written as (π + α).

$$\mathrm{Cos}\ 240 = \cos\ (\pi + 60)$$

In the third quadrant, the cosine has a negative sign.

$$\cos\ (\pi + 60) = -\cos 60 = -1/2$$

(o) tan (−330)=
(p) sin (−225)=
(q) cot (−135)=
(r) tan (−240)=

H. Rate of Change of Dependent Quantities (Derivative-Integral)

Let's consider the quantities A and B that change depending on each other. Let the quantity A be a function of the quantity B. The difference between the two values that B takes is called the amount of change of the quantity B and is denoted by $\Delta B = B_2 - B_1$. Let the value of the quantity A with respect to B_1 be A_1 and its value with respect to B_2 be A_2. The difference $A_2 - A_1 = \Delta A$ is the amount of change in response to the change in ΔB. The ratio $\Delta A/\Delta B$ is called the rate of change of the quantity A with respect to B. If ΔB changes as small as desired from a selected value of B, ΔA also takes very small values and the ratio $\Delta A/\Delta B$ remains equal to a certain number, the rate of change is called the derivative of the quantity a with respect to B and is denoted by dA/dB.

The derivative, as the limiting value of the rate of change, indicates the rapidity of change of the relative variable relative to the free variable. If we want to express it mathematically,

$$A \subset R, f : A \rightarrow R$$

$$\lim_{x \to a} \frac{f(x) - f(a)}{x - a} = f'(a) = \frac{d_f}{d_x}(a) = dfa$$

$$h \to 0 \text{ for } x - a = n \Rightarrow x \to a$$

$$\lim_{h \to 0} \frac{f(a + h) - f(a)}{h}$$

For example, in the left ventricular pressure change curve indicated in Fig. 1.25, the left ventricular pressure value increased from the P_0 value to the P_m value as a result of heart contraction in the time interval determined by T_1. In the T_3 time interval, the pressure decreases from the P_m value to the P_0 value. It is quite difficult to determine the T_1, T_2, and T_3 values exactly on this curve and especially to determine the fastest transition moment from the P_0 value to the P_m value and the fastest transition moments and steepness from the P_m value to the P_0 value with great accuracy. Let's examine an example of the derivative change values of the same left ventricular pressure change curve taken with a derivative tool and plotted together with P on a second channel on the graph (Fig. 1.26).

The dP/dt derivative values provide more detailed information about the rate of change and duration of P pressure. At moment X, the left ventricular pressure increase began, passed through the fastest increase value at moment U, and continued to increase with a gradually slowing increase rate in the UY interval, reaching the pressure value P_m, which is considered to be constant. Since it always remains at approximately zero value in the YZ interval, the pressure remains

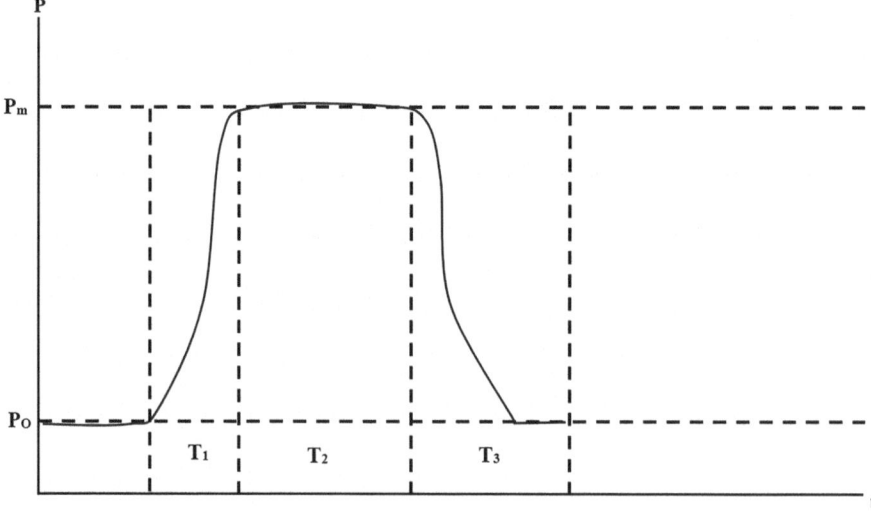

Fig. 1.25 Left ventricular pressure time change

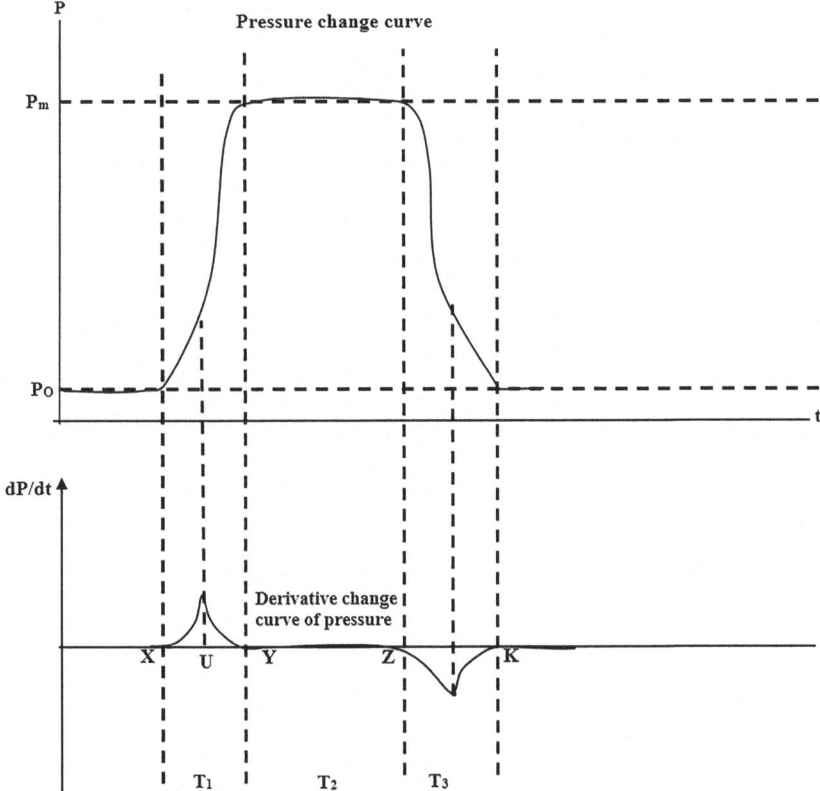

Fig. 1.26 Pressure variation curve and pressure derivative variation curve

approximately constant in this interval. Starting from moment Z, the derivative increases with the old value. Starting from moment Z, the ventricular pressure begins to decrease and, at moment V, passed through the fastest decrease value, and reached the P_0 value at moment K.

Some important derivation rules:

$$f(x) = x^n \Rightarrow f'(x) = n.x^{n-1}$$

$$(c.f(x))' = c.f'(x)$$

$$[f(x) \mp g(x)]' = f'(x) \mp g'(x)$$

$$[f(x) \cdot g(x)]' = f'(x) \cdot g(x) + g'(x)f(x)$$

$$\left[\frac{f(x)}{g(x)}\right]' = \frac{f'(x) \cdot g(x) - f(x) \cdot g'(x)}{[g(x)]^2}$$

$$f(x) = \sin x \Rightarrow f'(x) = \cos x$$

$$f(u) = \sin u \Rightarrow f'(u) = u' \cdot \cos u$$

$$f(u) = \sin^m u \Rightarrow f'(u) = m \cdot \sin^{m-1} u \cdot u' \cdot \cos u$$

$$f(x) = \cos x \Rightarrow f'(x) = -\sin x$$

$$f(x) = \tan x \Rightarrow f'(x) = \frac{1}{\cos^2 x} = 1 + \tan^2 x = \sec^2 x$$

$$f(x) = \cot x \Rightarrow f'(x) = -\frac{1}{\sin^2 x} = -1(1 + \cot^2 x) = -\csc^2 x$$

$$f(u) = \sqrt{u} \Rightarrow f'(u) = \frac{u'}{2\sqrt{u}}$$

$$f(u) = \arcsin u \Rightarrow f'(u) = \frac{u'}{\sqrt{1-u^2}}$$

$$f(u) = \arccos u \Rightarrow f'(u) = \frac{-u'}{\sqrt{1-u^2}}$$

$$f(u) = \arctan u \Rightarrow f'(u) = \frac{u'}{1+u^2}$$

$$f(x) = \log_a x \Rightarrow f'(x) = \frac{1}{x} \log_a e$$

$$f(x) = \ln x \Rightarrow f'(x) = \frac{1}{x}$$

$$f(u) = \ln u \Rightarrow f'(u) = \frac{u'}{u}$$

$$f(x) = a^x \Rightarrow f'(x) = a^x \cdot \ln a$$

$$f(u) = a^u \Rightarrow f'(u) = a^u \cdot u' \cdot \ln a$$

$$(e^u)' = u' \cdot e^u$$

When considering the two-variable function $z = f(x,y)$, if one of the two variables, for example, y, is given a constant value, a one-variable function is obtained with variable x. When the derivative of the z function with respect to the x variable is defined, this derivative is denoted as

$$z'_x, \frac{d_z}{d_x}, f'_x(x,y), \frac{d_f(x,y)}{d_x}$$

Similarly, the partial derivative with respect to y is defined as

$$z'_y, \frac{dz}{dy}, f'_y(x,y), \frac{d_f(x,y)}{d_y}$$

The Maclaurin and Taylor expansions of a function are generally given below:

$$f(x) = f(0) + \left(\frac{f'(0)}{1!}\right) x + \left(\frac{f''(0)}{2!}\right) x^2 + \frac{f'''(0)}{3!}) x^3 + \dots + \frac{f^{n-1}(0)}{(n-1)!} x^{n-1} + \dots$$

$$f(x) = f(a) + \ + \left(\frac{f'(a)}{1!}\right) (x-a) + \left(\frac{f''(a)}{2!}\right) (x-a)^2 + \frac{f'''(a)}{3!}) (x-a)^3 + \dots + \frac{f^{n-1}(a)}{(n-1)!} (x-a)^{n-1} + \dots$$

If we know or can determine the derivative of the function and one or more points belonging to the function, we can find the function itself. This process is called the concept of integral.

It has two different forms: indefinite and definite integral.

Some important integral rules:

$$\int f(x)dx = F(x) = \frac{d}{dx} F(x) = f(x) + c$$

$$\int af(x)dx = a \int f(x)dx$$

$$\int x^n dx = \frac{x^{n+1}}{n+1} + c(n \neq -1)$$

$$\int \frac{dx}{x} = \ln x + c$$

$$\int e^x dx = e^x + c$$

$$\int a^x dx = \frac{a^x}{\ln a} + c(a > 0)$$

$$\int \sin x dx = -\cos x + c$$

$$\int \cos x dx = \sin x + c$$

$$\int \frac{dx}{1+x^2} = \arctan x + c$$

$$\int \frac{dx}{\sqrt{1-x^2}} = \arcsin x + c$$

$$\int \frac{dx}{\sin^2 x} = -\cot x + c$$

$$\int \frac{dx}{\cos^2 x} = \tan x + c$$

Integration methods:

1. **Variable change method:**

$$\int \frac{du}{u} = \ln u + c \qquad\qquad \int u^m \cdot du = \frac{U^{m+1}}{m+1} + c$$

$$\int \frac{du}{1+u^2} = \arctan u + c$$

2. **Partial integration method:**

$$\int u \cdot dv = u^1 \cdot v - \int v \cdot du$$

3. **Simple fraction separation method:**

$$\frac{P_{(x)}}{Q(x)} = B(x) + \frac{k(x)}{Q(x)}$$

The difference between the definite integral and the indefinite integral concept is that the operation is only performed for certain limits.

$$\int_a^b f(x)dx = F(b) - F(a)$$

The most important areas of application of the definite integral are the ways it is used in area and volume calculations.

Area calculation:

$$X = a, x = b \text{ and } y = 0$$

$$A = \int_a^b f(x)dx$$

Volume calculation:

$$Y = f(x) \qquad x = a, \qquad x = b, \qquad y = 0$$

$$v = \pi \int_a^b y^2 dx$$

I. Multiple Exchange Relationship

In some cases, a quantity can change depending on more than one quantity. For example, gas pressure changes depending on both volume and temperature. (P=f (V), P=f(T)). We briefly show these two changes as P=f(V,T) and say that pressure is a function of volume and temperature. The type of relationship we encounter in biological systems is usually a relationship of change depending on many variables. For example, the amount of blood sent by the heart in a unit time (D) depends mainly on the blood volume in a beat (V), the number of beats in a unit time (f), and the aortic pressure (p). We can show this multivariate relationship with symbols as D=D (V,f,p).

Questions 9

1. If the derivative function expressing the pressure-time change in the left ventricle of a horse is y^2, what is the volume between the lines y=1, y=2, and x=0?

J. Exponential Function

The function $y = a^x$, where a > 0 is a number, is called an exponential function. Its graph is as shown in Fig. 1.27.

The most commonly used exponential functions in physics are those written as $U = A\,e^{at}$ or $V = B\,e^{-bt}$. In these expressions, A, a, B, and b are constants. The letter e is a constant number that represents the number 2.718 and is the base of the natural logarithm called the number e.

Use the ruler to find the values of the e^x and e^{-x} functions for various values of x (Table 1.10).

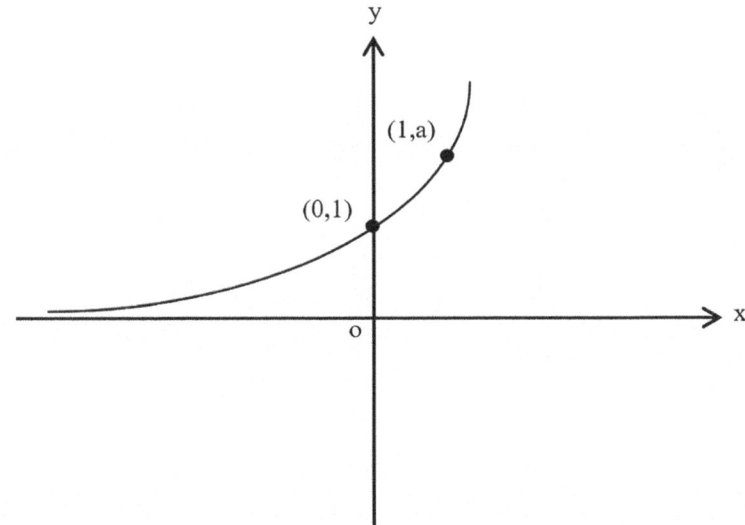

Fig. 1.27 Exponential function

Table 1.10 Exponential function e^x and e^{-x} values

x	0	0.2	0.4	0.6	0.8	1	1.2	1.4	1.8	2	2.2	2.4	2.6	2.8
e^x	1	1.22	1.49	1.82	2.33	2.72	3.32	4.05	6.05	7.39	9.02	11.02	13.46	16.44
e^{-x}	1	0.82	0.67	0.55	0.45	0.37	0.3	0.25	0.16	0.13	0.11	0.9	0.07	0.06

It is known that in ultrasound treatment, the intensity quantity of sound I varies depending on the depth as $I = I_0 e^{-\alpha x}$. Here, x is the depth, α is a constant number, and I_0 is the I value for $x = 0$. α is the absorption coefficient of the medium.

Questions 10

1. In ultrasound treatment, it is known that a wave with an intensity of 100 units initially affects various tissues, each of which is 4 cm thick. Calculate the radiation intensity at the end of each tissue separately. ($\alpha_{muscle} = 0.3$ cm^{-1}, $\alpha_{fat} = 0.2$ cm^{-1}, $\alpha_{bone} = 1$ cm^{-1})

$$I_{muscle} = 100 \, e^{-0,3 \,\times\, 4} = 100 \, e^{-1,2} = 100 \,\times\, 0.30 = 30 \text{ W/cm}^2$$
$$I_{fat} = 100 \, e^{-0,2 \,\times\, 4} = 100 \, e^{-0,8} = 100 \,\times\, 0.45 = 45 \text{ W/cm}^2$$
$$I_{bone} = 100 \, e^{-1 \,\times\, 4} = 100 \, e^{-4} = 100 \,\times\, 0.02 = 2 \text{ W/cm}^2$$

2. In ultrasound treatment, it is known that a wave with an intensity of 100 W/cm^2 initially affects a group consisting of muscle, fat, and bone tissues, each of which is 0.04 m thick. If there is a 50% absorbing medium in the middle of the bone that is so thin that its thickness can be neglected, find the intensities affecting the upper surface and the lower surface of the bone. ($\alpha_{muscle} = 0.3$ cm^{-1}, $\alpha_{fat} = 0.2$ cm^{-1}, $\alpha_{bone} = 1$ cm^{-1})
   First of all, it is necessary to convert the thickness of the tissues into the absorption coefficient unit of the medium.

$$0.04 \text{ m} = 4 \text{ cm}$$

$$I_0 = 100 \text{ W/cm}^2 \qquad I = I_0 \, e^{-\alpha x}$$

   Let's assume that it starts to affect the tissues in order. We need to look at the e^{-x} order of the table during the calculations.

$$I_{muscle} = 100 \, e^{-0.3 \,\times\, 4} = 100 \, e^{-1,2} = 100 \,\times\, 0.30 = 30 \text{ W/cm}^2$$
$$I_{fat} = 30 \, e^{-0.2 \,\times\, 4} = 30 \, e^{-0.8} = 30 \,\times\, 0.45 = 13.5 \text{ W/cm}^2$$
$$I_{upper \, surface \, of \, the \, bone} = 13.5 \text{ W/cm}^2$$
$$I_{absorber \, medium \, upper \, surface} = 13.5 \, e^{-1 \,\times\, 2} = 13.5 \, e^{-2} = 13.5 \,\times 0.13 = 1.755 \text{ W/cm}^2$$
$$1.755 \,\times\, 40/100 = 0.702 \text{ W/cm}^2$$
$$I_{lower \, surface \, of \, bone} = 0.702 \, e^{-1 \, x \, 2} = 0.702 \, e^{-2} = 0.702 \,\times\, 0.13 = 0,091 \text{ W/ cm}^2$$

3. In ultrasound treatment, it is known that a wave with an intensity of 4 W/cm^2 initially affects a group consisting of fat, muscle, and bone tissues, each of which is 2 cm thick. Find the intensity affecting the lower surface of the bone. ($\alpha_{muscle} = 0.3$ cm^{-1}, $\alpha_{fat} = 0.2$ cm^{-1}, $\alpha_{bone} = 1$ cm^{-1})

$$I_0 = 4 \text{ W/cm}^2 \qquad I = I_0 \, e^{-\alpha x}$$

Let's assume that it starts to affect the tissues in order. We need to look at the e^{-x} order of the table during the calculations.

$$I_{\text{muscle}} =$$

$$I_{\text{fat}} =$$

$$I_{\text{lower surface of bone}} =$$

4. It is known that in ultrasound treatment, a wave with an intensity of 8 W/cm^2 initially affects a group consisting of fat, muscle, and bone tissues, each of which is 2 cm thick. Find the intensity affecting the upper surface of the bone. ($\alpha_{\text{muscle}} = 0.3$ cm^{-1}, $\alpha_{\text{fat}} = 0.2$ cm^{-1}, $\alpha_{\text{bone}} = 1$ cm^{-1})

$$I_0 = 8 \text{ W/cm}^2 \qquad I = I_0 \, e^{-\alpha x}$$

Let's assume that it starts to affect the tissues in order. We need to look at the e^{-x} order of the table during the calculations.

$$I_{\text{muscle}} =$$

$$I_{\text{fat}} =$$

$$I_{\text{upper surface of the bone}} =$$

5. Calculate the radiation intensity on the lower surface of the bone of a 100-unit wave externally affecting a group of 4 cm thick muscle, fat, and bone tissues in the presence of a pin containing 40% absorber medium (thickness is neglected).

$$\left(\alpha_{\text{muscle}} = 0.3 \text{ cm}^{-1}, \alpha_{\text{fat}} = 0.2 \text{ cm}^{-1}, \alpha_{\text{bone}} = 1 \text{ cm}^{-1} \right)$$

$$I_0 = 100 \text{ W/cm}^2 \qquad I = I_0 \, e^{-\alpha x}$$

Let's assume that it starts to affect the tissues in order. We need to look at the e^{-x} order of the table during the calculations.

$$I_{\text{muscle}} =$$

$$I_{\text{fat}} =$$

$$I_{\text{upper surface of the bone}} =$$

$$I_{\text{absorber medium upper surface}} =$$

$$\text{........} \times 40/100 =$$

$$I_{\text{lower surface of bone}} =$$

K. Logarithmic Function

Let us consider a real number a with a value greater than one.

The inverse function of the function $x = a^y$ is shown as $y = \log_a x$ and is defined as the logarithm of x to the base a. The graphs are given in Fig. 1.28.

If the base is 10, it is not written, and if it is e, ln is used instead of $\log_e$. Some simple logarithm relations are given below:

1. $n = \log_a m \leftrightarrow m = a^n$
2. $\log_a (A. B) = \log_a A + \log_a B$
3. $\log_a (A/B) = \log_a A - \log_a B$
4. $\log_a (\alpha A) = \alpha \log_a A$
5. $\log_a (1/A) = - \log_a A$
6. $\log_a 1 = 0$
7. $\log_a a = 1$
8. $\log_a A^n = n. \log_a A$
9. $\log_a \sqrt[n]{A} = 1/n \, \log_a A$

For example, when the relationship between brain weight (E) and body weight (P) was examined, it was seen that fish and reptiles were grouped in one group and mammals in a separate group when different species (including fossils) were plotted on a logarithmic graph paper. From the examination of the data belonging to both

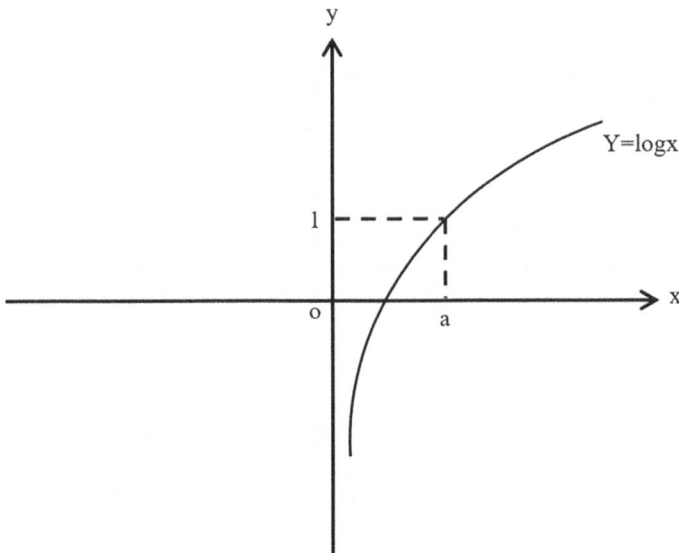

Fig. 1.28 Logarithmic function

groups, it was determined that there was a relationship between brain and body weights, approximately,

$$\log E = \log k + (2/3) \log P \text{ or } E = kP^{2/3}$$

Log k or just k in the relations is the degree of development of the brain, and it is expected that more intelligent animals have larger brains.

Another example of the use of logarithmic expression is the Weber-Fechner law. It is important under what conditions we can distinguish the difference between two quantities that we can sense through our senses. For example, although we can easily distinguish a length of 1 cm from a length of 2 cm with our observations, we cannot distinguish a length of 100 cm from a length of 101 cm, whose lengths are the same, without measuring and matching their ends. Based on such observations, Weber argued in the last century that in comparing quantities of the same kind, ratios are more important than differences and that if the difference between two stimuli is dI, the ratio $dI/I = K$ will be a constant depending on the type of stimulus. With similar thoughts, Fechner defined a concept called the perception factor as $dI^* = k \, dI/I$, by integrating which, the,

$I^* = k \log I/ I_0$ relation is obtained.

According to this expression, known as the Weber-Fechner law, the sensory or psychophysical intensity I^* should be proportional to the logarithm of the physical intensity (I) of the stimulus. In the relation, I_0 represents the threshold intensity of the stimulus, and k represents a constant related to the sensory type.

In Stevens' power law, it is stated that the psychophysical intensity sensation is proportional to a power of the physical stimulus intensity.

$$I^* = K \, I^n \qquad I^* = k'(I/I_0)^n$$
$$\log I^* = n \log I + \log K$$

Another place where logarithmic expressions are used is the concept of pH.

$$PH = -\log[H^+]$$
$$[H^+] + [OH^-] = 1.10^{-14}$$
$$POH = -\log[OH^-]$$
$$(-\log[H^+]) + (-\log[OH^-]) = -\log 1.10^{-14}$$
$$pH + pOH = 14 \qquad \text{Neutral is found as pH} = 7 \text{ and pOH} = 7$$

In our medicine, pH values are interpreted in relation to blood, rumen fluid (cows have four compartments in their fore stomachs, called rumen, reticulum, omasum, and abomasum) and urine.

Blood HCO_3^- concentration is regulated by the kidneys, while CO_2 concentration is regulated by the lungs. The ratio of kidney function to lung function is the blood pH value. Normal blood pH value is between 7.3 and 7.5. If the CO_2 concentration in the blood increases, respiratory acidosis occurs, and if the concentration decreases, respiratory alkalosis occurs. If the blood pH value decreases, metabolic acidosis occurs, and if the blood pH value increases, metabolic alkalosis occurs.

The pH value of rumen fluid is between 6.2 and 7.2. In rumen acidosis, this value drops to 5 and below, while a value between 7.5 and 8 suggests rumen alkalosis, and a value between 8 and 8.5 suggests putrefaction. If rumen fluid mixes with saliva, the pH value increases. There are small organisms called infusoria in rumen fluid. In acidosis, these organisms completely disappear, while when the pH value exceeds 7, the movements of these organisms decrease.

The urine pH values of cattle, sheep, and goats (ruminants) are alkaline while meat-eating dogs and cats (carnivores) are acidic. Ruminants and horses are fed with foods high in carbonate and phosphate ions, so their urine is alkaline, while meat-eating animals fed with carbohydrate-rich foods have alkaline urine. Pigs will have acidic urine if they are fed a high-protein diet and alkaline urine if they are fed a high-carbohydrate diet.

Questions 11

1. Calculate the blood pH for a dog with a blood HCO_3^- concentration of 20 mEqL and a CO_2 concentration of 5 mEqL.

Blood HCO_3^- concentration is regulated by the kidneys, while CO_2 concentration is regulated by the lungs. The ratio of kidney function to lung function gives the blood pH value. Therefore, the

$$pH = 20/5 = 4 \text{ value is reached.}$$

2.

$$pH_A = 7,5$$
$$pH_B = 4,5$$
$$pH_C = 8,5$$
$$pH_D = 7$$

The rumen fluid pH findings of cows A, B, C, and D are given above. Answer the following questions accordingly.

(a) Evaluate the health status of the cows by looking at the rumen fluid pH values.
   A. Alkalosis
   B. Acidosis
   C. Putrefaction
   D. Normal

(b) What are the infusoria like in the rumen fluid of cow B?

In cow B, there are no infusoria because the pH value of rumen fluid is 4.5.

(c) Draw the graph between infusoria motility and rumen fluid pH value in case of alkalosis.
(d) If no pathological findings were found in the clinical examination of cow C, what could be the reason for the increase in rumen fluid pH?

When rumen fluid mixes with saliva, the pH value increases.

3.

$$pH_1 = 8$$
$$pH_2 = 6$$
$$pH_3 = 5,5$$
$$pH_4 = 8,5$$

The urine pH values of various animals (horse, cat, dog fed carbohydrate-rich diet, pig fed protein-rich diet) are given above.

(a) Considering that the nutritional characteristics of these healthy animals are known, think about how the possible matches between the subject numbers and the animal species could be.
Horse.............. Number 1 or 4 (urine alkaline)
Cat............2 or 3 numbers (acidic urine)
Pig........2 or 3 numbers (acidic urine and a high protein diet)
Dog........1 or 4 numbers (alkaline urine and a high carbohydrate diet)
(b) What can be attributed to the alkalinity of horse urine?
It is related to excess carbonate and phosphate in the diet.
(c) If number 3 is pig, what is its pOH value?
POH = 14 − 5.5 = 8.5

4. Arrange the relationship between brain-body weights to indicate the L^2 weight and L^3 volume of an animal in terms of length L.

When the relationship between brain weight (E) and body weight (P) is examined, the relationship between brain and body weights is $E = k\, P^{2/3}$. If the operation is performed,

$$E = k\, P^{2/3} = k\, \left(L^2\right)^{2/3} = k\, L^{4/3}$$

5. Organize the relationship between brain and body weights to indicate the external surface area of an animal in terms of length L, L^2, and the density as $1/L^3$.

It is known that $d = m \times V$ and $E = kP^{2/3}$. Accordingly;

6. It has been determined that the anabolism activity of a 3 cm long marine animal with a square body surface is 85 units. Since it is observed that this creature grows in unit time, which of the following <u>cannot</u> be the value of β?
   (a) 4   (b) 3   (c) 2   (d) 1

7. Calculate the blood pH value for a dog with a blood HCO_3^- concentration of 20 mEqL and a CO_2 concentration of 5 mEqL.

Temperature and Heat

2

Abstract

In this section, the biological importance of body temperature, pulse, respiratory rate, mucosal examination, and elasticity concepts known as vital signs, especially temperature, is evaluated, with an emphasis on fever and its types. Information about thermometers is given, and the relationships between thermoregulation, thermotherapy, cauterization, and sunlight and the basic areas of therapy were emphasized. The Wien Shift Law and radiation band are examined, and the wavelengths of larger and smaller rays outside the visible region, which are biologically important, were examined. In particular, the hydrogen series are classified and the concepts of brightness and luminosity are explained. Information was given about air therapy, another method used in veterinary medicine. At the end of each section, the subject is detailed with explanatory questions.

Keywords

Body temperature · Pulse · Fever · Thermometers

2.1 Thermometers

There are various types of instruments used to measure temperature. The most commonly used ones are liquid thermometers. These degrees consist of a temperature-sensitive chamber filled with liquid and a capillary tube connected to it. In order to rate a thermometer, the melting temperature of ice and the temperature of boiling water under normal conditions are generally taken as two constant values. The most important thermometers arranged accordingly and their conversion formulas are given below (Fig. 2.1).

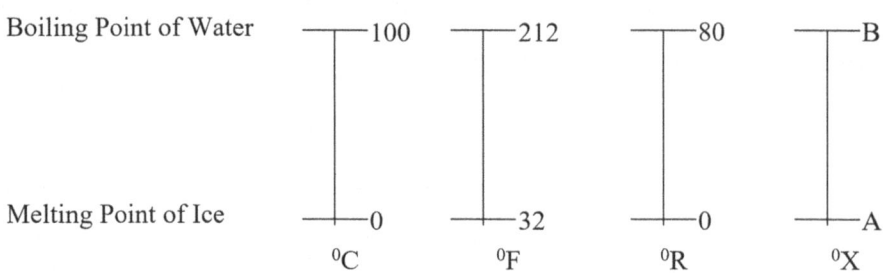

Boiling Point of Water   ─┬─100   ─┬─212   ─┬─80   ─┬─B

Melting Point of Ice   ─┴─0   ─┴─32   ─┴─0   ─┴─A
                                 ⁰C   ⁰F   ⁰R   ⁰X

Fig. 2.1 Graphs showing boiling and freezing points of water on various thermometers

$$C/100 = (F - 32)/180 = R/80 = (X - A)/(B - A)$$

Temperatures expressed on the Kelvin scale are called absolute temperatures and the following formula is used for conversion:

$$T\,(\text{Kelvin}) = t\,(^\circ C) + 273$$

Maximal thermometers are the most commonly used thermometers to determine temperature in living beings. The feature of these thermometers is that the mercury column does not fall down unless it is shaken properly. In addition, in recent years, thermometers that give results automatically or show color changes according to skin temperature have also been used.

2.2 Normal Temperature

In many sources, the concept of body temperature is encountered as an incorrect use. However, the unit of heat is calorie, which is the amount of heat required to increase the temperature of 1 g of water by 1 °C. The devices used to measure the amount of heat are called calorimeters. The normal temperature limits (°C) of animals depending on the species are given below.

Dog	37.5–39.0
Cat	38.0–39.5
Horse	37.5–38.0
Cow	37.5–39.5
Sheep	38.5–40.0

The body temperature of animals is taken primarily from the rectum, and rarely from the vagina for females. However, it should be noted that this value will be approximately 0.5–1.0 °C lower than the value taken from the rectum. Females have higher body temperatures than males. The same comparison can be made between pregnant and sterile animals and between young and old animals. When measuring temperature from the rectum, the mercury of the thermometer should be shaken to lower it to normal levels. Then it should be kept in the rectum for at least 5 min. During this period, it is necessary to ensure that the thermometer does not remain in the feces and that it definitely touches the mucosa. The thermometer should never be left in the rectum and another procedure should be performed. It should be checked whether there is a wound inside the rectum because in such a case, the locally increased temperature is considered as body temperature, which is wrong.

2.3   Fever (Febris)

Values above the normal body temperatures reported for various animal species are called fever or febrile. It can be examined in two different ways.

(a) **Septic fever:** Increased body temperature due to various disease agents entering the organism. We can classify the responsible agents as viruses, bacteria, fungi, and parasites (especially protozoa).
(b) **Aseptic fever:** Increase in body temperature due to foreign proteins or protein residues and necrotic tissue residues entering the organism, in addition to the factors causing septic fever.

2.4   Fever Types

As a result of observing the differences between morning and evening fevers, various types of fever were found. By using these values, it is easy to comment on some diseases.

(a) **Febris continua:** The differences between morning and evening fevers do not exceed one degree.
(b) **Febris remittens:** The difference between morning and evening fevers is more than one degree.
(c) **Febris intermittens:** As in the febris remittens type, daily differences are more than one degree, but unlike this, when the fever drops, it is at normal levels.
(d) **Febris recurrens:** It is a type in which fever and fever-free periods alternate at regular intervals.
(e) **Febris atypical:** A type of fever that does not resemble any other form.

Questions 12

1. At what temperature do Fahrenheit division thermometers and Réaumur division thermometers show the same number?

$$\frac{F-32}{180} = \frac{R}{80}$$

2. At what temperature does the number on the Fahrenheit section exceed the number on the Celsius section by 60?

$$F = C + 80 \qquad \frac{F - 32}{180} = \frac{C}{100}$$
$$C = 60 \ \degree C \quad \text{or} \ F = C + 80 \quad F = 140 \ \degree C$$

3. If a thermometer shows the melting point of ice as -30 and the temperature of boiling water as $+140$ under normal conditions, what value does it show as $40 \degree C$?

$$\frac{X - A}{B - A} = \frac{C}{100}$$
$$\frac{X - (-30)}{140 - (-30)} = \frac{40}{100} \qquad X = 38 \ \degree C$$

4. If a thermometer shows the temperature of boiling water as $+160$ under normal conditions and $37 \degree C$ shows it as $34 \degree$, what is the melting point of ice?

$$\frac{X - A}{B - A} = \frac{C}{100}$$
$$\frac{34 - A}{160 - A} = \frac{37}{100} \qquad A = -60 \ \degree C$$

5. At what temperature is the value indicated by the Kelvin thermometer 27 more than 4 times the value indicated by the $\degree C$ thermometer?
6. At what temperature do Fahrenheit division thermometers and Réaumur division thermometers show the same number?
7. At what temperature does the number on the Fahrenheit section exceed the number on the Celsius section by 60?
8. If a thermometer shows the melting point of ice as -20 and the temperature of boiling water as $+120$ under normal conditions, what does it show as $50 \degree C$?

2.5   Thermoregulation (Regulation of Body Temperature)

The combustion equation for organic matter in general is given below:

$$C_n H_{2n} O_n + n \, O_2 \rightarrow n \, CO_2 + n \, H_2O + \Delta H$$

The enthalpy expressed as ΔH can be positive or negative depending on the type of reaction. There is a balance between the formation of heat and the loss of heat in the organism. The formation of heat depends on the use of heaters such as the sun or the radiator, which provide heat from outside, or on the burning of the food taken. Heat loss occurs in the form of evaporation or radiation, depending on respiration and sweating. Thermoregulation is mainly controlled by the central nervous system (hypothalamus). The food taken into the organism is mainly burned by the muscles. If the environment is cold, the entire vascular system begins to tighten in order to prevent heat loss, and as a result, blood flow slows down. On the contrary, if the environment is hot, the vascular system relaxes and there is excess blood flow to the region. In both cases, the skin and hypothalamic receptors come into play, causing responses such as shivering or sweating. The behavioral equation and the feedback mechanism of the system examined in the previous section are seen here again. Studies have shown that proteins will undergo structural deterioration as a result of excessive increases (44–45 °C) in the normal body temperature (37 °C) in humans and that at low temperatures (27–28 °C), the circulatory system will be affected and the heart will stop. In our medical practice, if the pulse rate increases when the body temperature falls below normal limits, this is called the Death Cross and is a sign of collapse and death. In cat and dog practice, hot water thermometers are used to increase the body temperature of these patients. The normal pulse rates (/min) of various animal species are given below:

Dog	70–120
Cat	110–130
Horse	28–40
Cow	60–80
Sheep	70–80

In horses, the pulse is taken from the inner surface of the A. maxillaris externa, or in other words, the internal A. maxillaris interna. In cows, A. maxillaris externa is used to take pulse. In both species, this artery passes through the incisura vasorum of the mandible bone. In addition, a pulse can easily be felt from A. coccygea on the lower surface of the tail root of cows. In cats and dogs, A. femoralis, which runs on the inner surface of the hind leg, is used to take the pulse. Although sheep and goats are ruminants, the pulse is taken from A. femoralis, just like in carnivores. When measuring the pulse, we must hold the area with all fingers of our hand because we cannot know which finger we will feel the pulse with. In addition, we should neither apply too much pressure nor hold too lightly. In the latter case, the pulse may not be felt, while in the former, we can take our own pulse.

In horses, there are two more important arteries called A. metacarpalis digitalis and A. metatarsalis digitalis running in the front and hind legs. Under normal conditions, no pulse can be felt from these arteries. However, in cases of fourbure disease, feeling a pulse from these arteries is a pathognomonic (specific for that disease) symptom for this disease.

2.6   Thermotherapie (Heat Healing)

The therapeutic effects of heat have been used in various areas of our medicine. Studies have shown that mild heat applications reduce pain and decrease muscle tone. Conversely, high heat applications have shown that pain gradually increases and muscle tone and peristalsis increase. While it has been reported that long-term heat applications cause muscle relaxation and fatigue, short-term heat applications have a stimulating effect.

2.7   Cauterization

It is a treatment method performed with a very hot tipped stick or special tools. This method, based on burning, is used in tumor treatment and to stop bleeding.

2.8   Heliotherapy (Sunlight Healing)

The UBVRI luminance system is divided into five main classes depending on the wavelengths of the bands (Fig. 2.2):

(a) Ultraviolet (UV): Ultraviolet light
(b) Blue (B): Blue light
(c) Visual (V): Visible light
(d) Red (R): Red light
(e) Infrared (I): Infrared light

2.8.1   Wien's Shift Law

Color is a function of temperature. As temperature increases, wavelength shortens. For example, objects with short wavelengths appear white or white-blue, while the color of the very hot sun shifts to yellow, and objects with cold colors shift to red because in this region, wavelength increases to 1200 nm while temperature decreases.

The use of ultraviolet rays in medicine is related to their short wavelengths and their ability to enter the body at a higher rate. These rays have a high chemical startle effect. Studies have shown that anemic individuals are exposed to these rays,

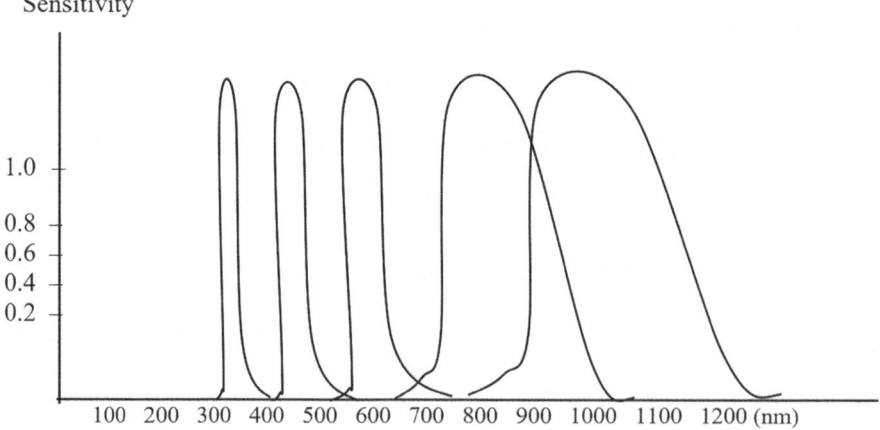

Fig. 2.2 Radiation band

resulting in an increase in their leukocyte and hemoglobin values. However, no change occurs in healthy individuals. It has been reported that the leukocyte count increases in horses and dogs, especially within the first 24 h, and returns to normal within 3 days. For this reason, such an application is made in order to create resistance against infections. In addition, ultraviolet lights are used in the treatment of diseases such as eczema, alopecia, and rickets and in the sterilization of the operating room. In the case of rickets, which occurs especially in young animals due to Ca and P imbalances, Quartz-Hg or Quartz-Cd bulbs are used in the cabins created. It is used in addition to the treatment for 5–30 min from a distance of approximately 0.5–1 m, as long as the doctor recommends. It is of great benefit for patients to benefit from sunlight in accordance with the season. This treatment is also used to activate the immune system in the treatment of eczema, one of the important skin diseases of dogs.

Treatment with sunlight is actually a phototherapy. Of course, it is also known that sunlight has many undesirable effects. This type of skin diseases and tumors are very common especially in gray-haired horses, short-haired dogs, and cats. In addition, there are plants in nature that contain some photodynamic substances. The most important of these are plants such as Hypericum perforatum (perforated fishbone grass), Fagopyrum esculentum (buckwheat), Trifolium alba (capricorn species), and chemical substances such as carbon tetrachloride and phenothiazine. When these plants are consumed by ruminants, horses, and pigs, if the animals are also exposed to sunlight, they cause a dermatitis called dermatitis solaris.

Apart from the Wien Shift Law, another law that expresses the energy distribution of a source with a certain temperature depending on the wavelength is Planck's law. The distribution of radiation according to wavelength changes as a function of temperature. The ability of an object to absorb only certain wavelengths of the radiation energy coming to it is called spectral absorption ability. A blackbody is defined as an object that does not reflect the radiation falling on it, absorbs all of it, and then reradiates it after a certain period of time. In fact, it is impossible to find such an object in nature. In blackbody radiation, energy increases as the temperature increases. Energy is proportional to the fourth power of temperature.

$$E = \sigma T^4$$

The relation expressing Planck's law is as follows:

$$B(\lambda_1 T) = \frac{2hc^2}{\lambda^5} \frac{1}{e^{\frac{hc}{\lambda kT}} - 1}$$

During the transition from one energy level to another, electromagnetic radiation is emitted. When the atom's energy changes by ΔE, the atom emits an electromagnetic beam called a photon, which has a given frequency.

$$\Delta E = h\,\gamma$$
$$\gamma = c/\lambda$$
$$\overline{v} = \frac{1}{\lambda} = R\left(\frac{1}{n_1^2} - \frac{1}{n_2^2}\right)$$

We can organize the frequency or wavelength formula depending on the different series of the hydrogen atom. The most commonly used series are the Lyman, Balmer, Paschen, Bracket, and Pfund series, respectively.

	n_1	n_2
Lyman series	1	2, 3, 4,...
Balmer series	2	3, 4, 5,...
Paschen series	3	4, 5, 6,...
Bracket series	4	5, 6, 7,...
Pfund series	5	6, 7, 8,...

Questions 13

1. How many ergs and eV does a photon with a wavelength of 1000 Å have?
2. When a photon hits a hydrogen atom, the electron in step 1 jumps to step 4. Write the relation that gives the wavelength of the photon.

$$1/\lambda = 109,677 \,(1/1^2 - 1/4^2) = 109,677 \,(1 - 1/16)$$

3. An electron in a hydrogen atom moves from the sixth step to the second step. Write the relation that gives the wavelength of the resulting photon.

$$1/\lambda = 109,677 \,(1/2^2 - 1/6^2) = 109,677 \,(1/4 - 1/36)$$

4. Find the relations giving the wavelengths of the second and fourth lines of the Balmer series.
   For Balmer series, n_1 is always taken as 2. Accordingly, the relations giving the second and fourth series are as follows:

$$1/\lambda = R\left(1/2^2 - 1/4^2\right) = R\,(1/4 - 1/16)$$
$$1/\lambda = R\left(1/2^2 - 1/6^2\right) = R\,(1/4 - 1/36\,)$$

5. Find the relations giving the wavelengths of the first and third lines of the bracket series.
   For bracket series, n_1 is always taken as 4. Accordingly, the relations giving the first and third series are as follows:

$$1/\lambda = R\left(1/4^2 - 1/5^2\right) = R\,(1/16 - 1/25)$$
$$1/\lambda = R\left(1/4^2 - 1/7^2\right) = R\,(1/16 - 1/49)$$

6. Find the relations giving the wavelengths of the second and fourth lines of the Pfund series.

   For Pfund series, n_1 is taken as continuous ...... Accordingly, the relations giving the first and fourth series are as follows:

   $1/\lambda = $ ....

   $1/\lambda = $ ....

7. Find the relations giving the wavelengths of the first and third lines of the Paschen series.

   For Paschen series, n_1 is taken as continuous .... Accordingly, the relations giving the second and third series are as follows:

   $1/\lambda = $ ....

   $1/\lambda = $ ....

8. An electron in a hydrogen atom falls from the sixth step to the third step, respectively. Write the relations that give the wavelengths of the photons that emerge.

   $1/\lambda = R\left(1/5^2 - 1/6^2\right) = R\left(1/25 - 1/36\right)$

   $1/\lambda = R\left(1/4^2 - 1/5^2\right) = R\left(1/16 - 1/25\right)$

   $1/\lambda = R\left(1/3^2 - 1/4^2\right) = R\left(1/9 - 1/16\right)$

9. Calculate the distance between the third and second lines in the Pfund series of hydrogen.

   $1/\lambda_2 = R\left(1/5^2 - 1/7^2\right) = R\left(1/25 - 1/49\right) = $ ....

   $1/\lambda_3 = R\left(1/5^2 - 1/8^2\right) = R\left(1/25 - 1/64\right) = $ ....

   $\Delta\lambda = $ ....

10. What is the minimum wavelength of photon that an electron in the fourth step of a hydrogen atom must absorb in order to move far enough away from the attraction?

11. Find the wavelength and energy of the photon that a hydrogen atom with an electron in the second digit must absorb in order to ionize (lose an electron).

12. The wavelength of the photon from the (4-1) transition in an atom is 4265 Å and the wavelength of the photon from the (4-3) transition is 8379 Å. Calculate the energy and wavelength of the photon from the (3-1) transition.

13. The wavelength of the photon emitted when an electron in a hydrogen atom moves from the fourth step to the first step is the maximum wavelength. Accordingly, find the luminosity by accepting the temperature and the radius of the photon as 400 m.

$1/\lambda = R \, (1/1^2 - 1/4^2) = 109677 \, (1/1 - 1/16) = \dots\dots\dots\dots$

Using the Wien Slip Law
Its luminosity is ....

14. Calculate the wave number, wavelength (Å) and energy (eV) corresponding to the Rydberg constant of hydrogen.
15. If the change in frequency of a photon of 2 Å colliding with an electron is 0.0001γ, calculate the angle at which the photon is scattered.
16. If it is assumed that the ionization energy is defined as the transition from the first level to the infinite level for each series, calculate the ionization energies (eV) for the Lyman, Balmer, Paschen, Brackett, and Pfund series of the hydrogen atom.
17. Draw the electromagnetic radiation band and find the value of the wavelengths (Å) for the first lines of all series of the hydrogen atom. Calculate the difference between the wavelengths you found according to the infinite level in the second question and write down which region they fall into.

Radiation intensity (I): The amount of energy coming from a unit area through a unit angle in a unit time.

Radiant power or luminousness (L): It is the total energy emitted in all directions and at all wavelengths per unit time.

$L = 4\pi R^2 \, \delta T^4$

Apparent brightness (m): The amount of energy coming to a unit area per second.

Absolute brightness (M): The apparent brightness of a source measured 10 parsecs away.

2.8.2 Brightnesses

1. Visual Brightness (m_v): The brightness measured by the eye is called visual brightness. The eye is sensitive to radiation at its maximum level when the wavelength is around 5500 A,
2. Photographic Brightness (m_{pg}): It is related to blue, violet, and ultraviolet radiations and is generally measured with photographic plates with a wavelength of 4300 A.
3. Photovisual Brightness (m_{pv}): Due to the variable eye sensitivity, visual brightness is measured with plates having a wavelength of 5550 A. m_{pv} brightness has replaced m_v brightness.

2.8.3 Color Index (Color Scale)

Color index, defined as the difference between photographic brightness and photovisual brightness, varies inversely with temperature.

$$C.S. = m_{pg} - m_{pv} = B - V$$

For example, around 11,000 °K C.S. $= 0$ is accepted, when the temperature rises above 11,000 °K, $m_{pg} < m_{pv}$, so C.S. will be evaluated as negative, when the temperature falls below 11,000 °K $m_{pg} > m_{pv}$, so C.S. will be evaluated as positive.

2.8.4 Color Excess

The difference between the color scale measured for any source and the color scale values previously known for the same type of sources is formulated as.

$$E_{B-V} = C.S.(observed) - C.S.(expected) = (B - V) - (B - V)_0$$

If the color residue is positive, it is an absorbing medium. In general, the brightness of the absorbing medium is taken as three times the color residue.

2.9 Aerotherapie (Air Treatment)

In our medicine, it is used in three different ways in the treatment of diseases related to the respiratory system. These are:

(a) Pneumotherapy: It is the use of air for therapeutic purposes. The most suitable temperature is around 12–16 °C. When it goes below 0 °C, chills and body temperature decreases and intestinal peristalsis increases; when it goes below − 10 °C, immobility and freezing are observed. When it goes above 20 °C, it has been determined that body resistance decreases and microorganisms multiply.
(b) Antimatrie (Incense): It is applied in the form of a room and bag method, especially in horses for respiratory system diseases. For the first method, a special shelter is prepared. The animal is tied here and allowed to breathe the medicated steam slowly. In the second method, a bag filled with straw is hung around the horse's neck. The medicine is applied to this grass and hot water is poured. The animal is allowed to be affected by the steam for 5–10 min.
(c) Climatotherapy: It is a form of treatment that takes advantage of climate changes, especially in humans.

Questions 14

1. The absolute brightness of a source 230 parsecs away is magnitude −1.

   (a) Calculate whether this source is visible to the naked eye or not.

   Distance (d) $= 230$ pc　　　Absolute Brightness (M) $= − 1$ magnitude
   　　The Apparent Magnitude (m) can be found using the formula
   $m − M = 5 \log d − 5$.

$m - (-1) = 5 \log 230 - 5$

$m + 1 =$

We can only see sources up to magnitude 6 with our eyes. Therefore, since m > 6, our source is visible to the eye.

(b) If there was a magnitude 1 absorption medium between us, what would be the apparent magnitude of the source?

Apparent Brightness (m) should be recalculated using the formula

$m - M = A + 5 \log d - 5$.

$m - (-1) = 1 + 5 \log 230 - 5$

$m + 1 =$

2. Which is closer, the first source with a parallax of 0″.049 or the second source with a distance of 96.49 pc?

For the first source $p'' = 0''.049$

$d_1 = 1/p''$ $d_1 = 1/0''.049 = \ldots\ldots\ldots$ pc

For the second source $d_2 = 96.49$ pc

Therefore, since $d_1 < d_2$, the first source is closer.

3. Can a source with a distance of 2250 pc and an absolute magnitude of -1.5 be seen with the naked eye?

Distance (d) = 2250 pc   Absolute Brightness (M) = -1.5 magnitude

Apparent Brightness (m) can be found using the formula

$m - M = 5 \log d - 5$

$m - \ldots\ldots = \ldots\ldots\ldots\ldots\ldots\ldots$

We can only see resources up to the sixth magnitude with our eyes. Therefore, $\ldots\ldots\ldots\ldots\ldots\ldots\ldots\ldots\ldots\ldots\ldots\ldots\ldots\ldots\ldots\ldots\ldots\ldots\ldots\ldots\ldots\ldots\ldots$

4. If the absolute brightness of a source is magnitude -3 and its distance is 900 pc, find its apparent brightness and parallax.

Distance (d) = 900 pc   Absolute Brightness (M) = -3 magnitude

Apparent Brightness (m) can be found using the formula

$m - M = 5 \log d - 5$

$m = \ldots\ldots\ldots\ldots\ldots\ldots\ldots$

For the first source = 0″.049

$p'' = 1/d$ $p'' = 1/900$ $= . \ldots\ldots\ldots$

5. The absolute brightness of a source 120 pc away is magnitude -1. Calculate whether the source can be seen or not. If there is an absorption of magnitude 1, what will be the apparent magnitude?

Distance (d) = 120 pc   Absolute Brightness (M) = -1 magnitude

The apparent magnitude (m) can be found using the formula

$m - M = 5 \log d - 5$

$m - (-1) = 5 \log 230 - 5$

$m + 1 =$

We can only see sources up to magnitude 6 with our eyes. Therefore, since m > 6, our source is visible to the eye.

If there was a magnitude 1 absorption medium between us, what would be the apparent magnitude of the source?

Apparent Brightness (m) should be recalculated using the formula

$m - M = A + 5 \log d - 5$

$m - (-1) = 1 + 5 \log 230 - 5$

$m + 1 = 6.5$

$m = 5.35$ magnitude

6. At what distance would a source with an apparent magnitude of 10 located 200 parsecs away have to appear as a magnitude 6 source?

Distance (d) = 200 pc   Apparent Brightness (m) = 10 magnitude

Absolute Luminance (M) can be found using the formula

$m - M = 5 \log d - 5$

$10 - M = 5 \log 200 - 5$

$10 - M = 6, 5$

$M = 3, 5$ magnitude

Since the source is the same source, we take the M value as the same in the second case and calculate the distance.

$6 - M = 5 \log d - 5$

$d = \ldots\ldots\ldots\ldots$

7. The apparent brightness of a source is 2.49 magnitude and its distance is 120 pc. How far away must the source be for its apparent brightness to fade by 0.1 magnitude?

Distance (d) = 120pc   Apparent Brightness (m_1) = 2.49 magnitude

If we use the formula $m - M = 5 \log d - 5$

$m_1 - M = 5 \log d_1 - 5$ relation is obtained.

If the same equation is written for a 0.1 magnitude dimming in apparent brightness,

$m_2 - M = 5 \log d_2 - 5$ Apparent Brightness (m_2) = 2.49 + 0.1 = 2.59 magnitude If two relations are subtracted from each other,

$m_1 - M - (m_2 - M) = 5$ $\log d_1 - 5 -$

$(5 \log d_2 - 5) m_1 - M - m_2 + M = 5 \log d_1 - 5 - 5 \log$ $d_2 +$

$5m_1 - m_2 = 5 \log d_1 - 5$ $\log d_2$

(Using the definition of the difference of logarithms) $m_1 - m_2 = 5$ $\log d_1 / d_2$

If the values are written in their places $\Delta \lambda = \ldots\ldots\ldots\ldots\ldots\ldots$

8. If the radius and absolute temperature of a light source are doubled, what will be the change in their luminosities?

9. The maximum and minimum apparent brightness of a source are measured as 3.0 and 6.0 magnitudes, and the maximum and minimum temperatures from its spectrum are found to be 2800 and 1600 K. Find the ratio of their radii in these two states.

10. Find the color indices of the sources given as α (m_{ph} = 3.6 magnitude, m_v = 2.4 magnitude), β (m_{ph} = 4.4 magnitude, m_v = 3.8 magnitude), and γ (m_{ph} = 7.6 magnitude, m_v = 5.4 magnitude). Determine the bluest and reddest.

$$(C.I.)_{\alpha} = m_{ph} - m_{pv} = 3,6 - 2,4 = 1,2 \text{ magnitude}$$
$$(C.I.)_{\beta} = m_{ph} - m_{pv} = 4,4 - 3,8 = 0,6 \text{ magnitude}$$
$$(C.I.)_{\gamma} = m_{ph} - m_{pv} = 7,6 - 8,1 = -0,5 \text{ magnitude}$$

The bluest one: ....
The reason: ....
The reddest one: ....
The reason: ....

11. The (B-V) value of a source with an apparent magnitude of 8.34 and an absolute magnitude of 1.20 has been measured as +1.35. If the (B-V) value of such sources is normally expected to be +1.15, calculate the distance of the source.

12. The density of the first of the two spherical sources is 8×10^{-2} times the other, and its mass is 23 times the other. The absolute brightness of the first source is −6 magnitude, and the second is +4 magnitude. How many times has the temperature of this source changed?

Wave Motion

3

Abstract

In this section, firstly, information about wave motion and wave equations and parameters are given and the differences between monochromatic waves, electromagnetic waves, and magnetic waves are evaluated. Newton's laws and Coulomb's law are explained, Gauss's theorem is examined, and various changes depending on the phase difference are examined. At the end of each section, the subject is tried to be detailed with explanatory questions.

Keywords

Waves · Newton · Coulomb · Gauss

3.1 Monochromatic Wave Equations

It is defined as a wave that spreads from one point to another in a substance or medium. We can examine waves in three classes:

(a) Transverse waves: The direction of the wave propagation and the directions of the vibrating particles are perpendicular to each other, for example, waves formed on the surface of still water.
(b) Longitudinal waves: The direction of the wave travel is the same as the direction of the vibrating particles, for example, sound waves.
(c) Elliptical waves: It is a type of wave that has a component perpendicular to the direction of wave propagation, for example, waves on the sea.

M. E. Or, *Medical Physics for Veterinary and Related Studies: An Introductory Textbook on Mathematical and Physical Principles*,
https://doi.org/10.1007/978-3-031-97355-0_3

There is no displacement of matter or medium in waves. The one-dimensional or monochromatic wave function can be expressed by one of the following equations:

$$\Psi (x, t) = A \sin k (x \pm v\, t)$$
$$\Psi (x, t) = A \sin (k\, x \pm w\, t)$$
$$\Psi (x, t) = A \sin 2\pi (x/\lambda \pm t/\tau)$$
$$\Psi (x, t) = A \sin 2\pi \nu (x/v \pm t)$$
$$\Psi (x, t) = A \sin 2\pi (\tilde{\upsilon} \pm \nu\, t)$$

If we define some concepts in these equations,

Frequency (ν): The unit is the number of vibrations per second and is expressed in Hertz
Amplitude (A): The point at which the wave has maximum displacement
Wavelength (λ): The distance between two successive waves
Period (τ): The repetition period of the wave in time
Angular frequency (w): Expression of frequency in radians per second
Wave speed (v): The speed at which a point on a wave moves
Propagation constant (k): $2\pi/\lambda$
Wave number ($\tilde{\nu}$): $1/\lambda$

3.1.1 Phase and Phase Speed

In the wave equation $\psi (x, t) = A \sin (kx - wt)$, the magnitude $(kx - wt)$ is called the phase of the wave. When $x = 0$, $t = 0$, $\psi = 0$ is a special case where the initial phase is assumed to be zero.

$$\psi(x, t) = A \sin (kx - wt + \varepsilon) \qquad \varepsilon : \text{Initial phase}$$

3.2 Electromagnetic Waves

Light is an electromagnetic wave with a frequency ranging from 400 to 700 nm. Electromagnetic waves propagate in space and carry energy. The electromagnetic radiation bands are specified in which electromagnetic waves are found in different wavelength ranges (Fig. 3.1).

The first electromagnetic waves are γ waves. These waves have the shortest wavelengths but the highest frequencies. Therefore, they have the highest energy. γ rays are formed by radioactive nuclear decays and can be harmful when they enter the living organism intensively.

X-rays are obtained by nuclear reactions. They are widely used in our medicine because they can easily pass through soft tissues and can be stopped in bones and other solid objects.

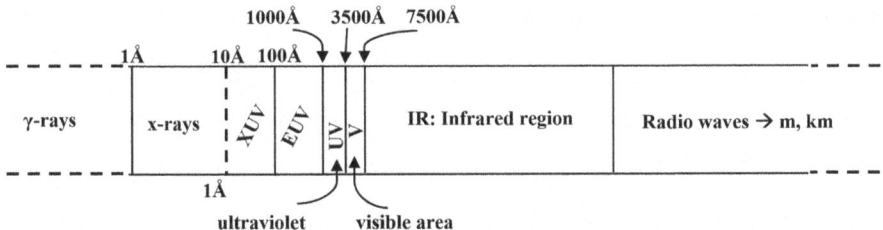

Fig. 3.1 Extended radiation band

The source of the UV band is the sun. The effects of these rays are used to tan. However, long-term interaction can lead to skin cancer. It is known that these rays are produced during the electron transitions of atoms.

Infrared rays are produced as a result of the vibration movements of molecules. Microwaves are waves used in telephone communication. They are waves related to the rotational movements of molecules. In recent years, their beneficial and harmful effects have become debatable as they have entered the food sector.

Radio waves, radio and television broadcasts are made with these waves. The most important radio wave sources in the universe are Pulsars.

In general, the equation of an electromagnetic wave is expressed as:

$$Ey\ (x, t) = E_{0y}\ \cos\ [w\ (t - x/c) + \varepsilon]$$

(E_{0y}): Amplitude of electromagnetic wave.

If it is desired to switch from the electromagnetic wave equation to the magnetic field equation, the new amplitude is obtained by simply dividing the amplitude by the speed of light and the equation is rewritten.

$$B_z = E_{0y}/c\ \cos\ [w\ (t - x/c) + \varepsilon]$$

$B_{0z} = (E_{0y})/c$: Amplitude of the wave in the magnetic field

Let's consider an object with a charge q_0 in space. The electric force F acting on this object is the electric field formed at this point,

$$E = F/q_0$$

The direction of the electric field is along the line passing through the charge. If q (+), it is away from the charge; if q(−), it is toward the charge.

Coulomb's law: The effect of two electric charges on each other is directly proportional to the charges and inversely proportional to the square of the distance between them.

When the electric field exerts a force F on a particle with a q charge, it will also create an acceleration. The system created by placing two equal but opposite q

Fig. 3.2 Electrical dipole system

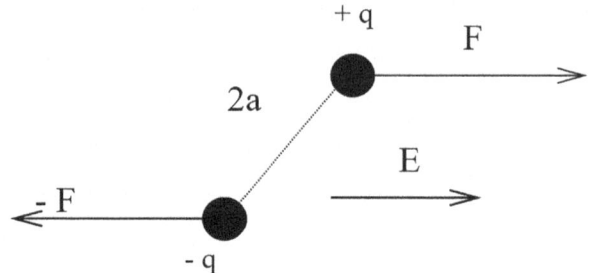

charges at a distance of 2a from each other is called the electric dipole (Fig. 3.2). It has been proven that the electric field at a point r away from the center of the dipole is

$$E = F/q_0 = (1/4\pi\ \epsilon_0)\ q/r^2 \qquad E = (1/4\pi\ \epsilon_0)\ 2.2aq/r^3$$

$$\text{for large values of the distance.}$$

The product of 2aq in the relation is called the electric dipole moment (p). It is a vector and its direction is from the negatively charged charge to the positively charged direction.

3.2.1   Newton's Laws

1. Unless an external force acts on an object, the object either remains at rest or moves in a uniform linear motion at a constant speed.
2. The rate of change in momentum of an object is proportional to the force acting on the object and is in the same direction.
3. The mutual effects of two bodies on each other are equal and opposite.

The angular momentum of an object with mass m while it is moving can be expressed with the location vector and the velocity vector as follows:

$$P = m\,\vec{v} \qquad \vec{L} = \vec{r}\,.m\,\vec{v}$$

$$\dot{\vec{L}} = \dot{\vec{r}}\,.m\,\vec{v} + \vec{r}\,.m\,\dot{\vec{v}}$$

$$\dot{\vec{L}} = \vec{r}\,.\vec{F} = \vec{\zeta}$$

3.2.2   Torque

The torque vector (τ) is known as the moment of force F.

$$\text{If } \tau = 2aF\sin\theta \text{ and } F = qE \text{ then } \tau = 2aqE\sin\theta = pE\sin\theta$$

Work is done by changing the direction of the electric dipole. If we rotate the dipole axis by θ, the work done is,

$$w = \int_{\theta_0}^{\theta} \zeta d\theta = \vec{u}$$

3.2.3 Gauss Theorem

The total amount of electric flux through an arbitrary closed surface enclosing one or more charge distributions is equal to the total charge trapped within the surface.

Questions 15

1. Find the frequency, wavelength, period, amplitude, and phase velocity in the wave equation given as $\psi = 1000 \sin \pi \, (3.10^6 \, x - 9.10^{14} \, t)$.
2. Let's consider the plane electromagnetic wave given by the expression $Ey = 2 \cos [2\pi. \, 10^{14} \, (t - x/c) + \pi/2]$. Write its frequency, amplitude, speed, propagation constant, and magnetic field equation.
3. Calculate how many cm away from each point the two charges $q_1 = 10^{-6} \, C$ and $q_2 = 2 \times 10^{-6} \, C$, which are 10 cm apart, must be in order for the electric field created by them to be zero.
4. A dipole formed by two-point charges with $q = \pm 10^{-6} \, C$ values 2 cm apart is placed in a field of $E = 10^5$ N/C. Calculate the maximum value of the torque exerted by the field on the dipole. With an external effect, the dipole is rotated 180° and brought into line with the field. Calculate the work done by the external forces.
5. Find the gravitational force between two 1 cubic meter charges that are 1 m apart.
6. It is known that the electrostatic force between two ions of the same type, 5×10^{-10} m apart, is 3.7×10^9 N. Calculate the charge value on each ion.
7. What should be the distance between two protons so that the repulsive force between them is equal to their weight on Earth? ($m_p = 1.7 \times 10^{-27}$ kg) ($q = 1.6 \times 10^{-19}$ C).
8. What is the magnitude of a point charge that creates an electric field intensity of 20 N/C at a point 50 cm away from itself?

9. If the total amount of electric flux passing through an arbitrary closed surface surrounding more than one charge distribution is given by the function $D = s^2 - 3\,s + 4$, calculate the total charge trapped in the surface for $s = 1$.

10. At a point r away from the dipole center on the side where the +q charged particle is located along the axis of the dipole given in the figure, the electric field is for the largest value of r, prove that $E = (1/2\,\pi\,\varepsilon_0)\,(p/r^3)$,

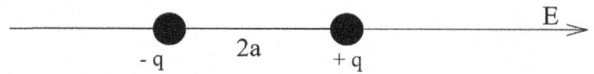

X-Rays

4

Abstract

In this section, information is given about the scattering mechanisms of X-rays and their differences are evaluated especially on atoms and ions. Again, how they are used for diagnosis and treatment purposes is examined by giving formulas in energy calculations and Compton scattering. In addition to the techniques used for image improvement, information is given about old X-ray baths and a comparison of digital and analog systems is made. Approaches are made based on fluoroscopy and angiography. At the end of each section, the subject is tried to be detailed with explanatory questions.

Keywords

X-ray · Compton scattering · Fluoroscopy · Angiography

4.1   Absorption Types of X-Rays

4.1.1   Classical-Unchanged Scattering

In this form of absorption, a low-energy X-ray photon hits an electron that orbits the nucleus but has very strong bonds and continues on its way in another direction without any loss of energy.

4.1.2   Photoelectric Effect

In this absorption form, the X-ray photon hits one of the electrons around the nucleus and gives all its energy to this electron, and this electron is removed from the atom and replaced. As a result of this event, the atom turns into an ion. Since the energy of the incoming photon is greater than the binding energy of the electrons, the photo-electric effect occurs.

4.1.3 Compton Scattering

In this absorption form, the X-ray photon hits one of the electrons in the orbit, but some of its energy is used to eject the electron from the atom. The other part creates a new photon. Here, during the first impact, there is a change in momentum and kinetic energy for the electron. The newly formed photon has less energy and a longer wavelength. As a result of this event, it is observed that the atom is ionized.

4.1.4 Pair Formation

A high-energy (at least 1.02 meV) X-ray photon hits the atomic nucleus, causing the formation of an electron and a positron particle, which itself disappears in the environment. In other words, energy is transformed into these substances. The positron collides with either the other electron formed or another electron in the environment and becomes two electromagnetic photons (with an energy of 0.51 meV). This event is known as annihilation radiation. In this mechanism, unlike the Compton and photoelectric events, the atom does not turn into an ion. In addition, a pair formation is never followed by another pair formation. However, photoelectric and Compton events can follow each other after the pair is formed. Which mechanism will work is related to the magnitude of the energy possessed.

In our medicine, the photoelectric effect and the Compton effect, which occur with low-energy photons, are mostly used for diagnosis, while the pair formation mechanism is used in treatment processes that require high energy. Since high-energy rays are more difficult to absorb, they enter the tissues more easily. Other factors affecting absorption (A) are density (d), thickness (t), wavelength (λ), and atomic number (Z). Accordingly,

$$A = Z^4 . \lambda^3 . d.t$$

Questions 16

1. If it is known that the absorption of a high-energy hydrogen photon on a 10-gram cube-shaped tissue with a thickness of 3 cm is 30, find the wavelength. (H = 1).

With the help of the relation $\qquad\qquad$ $A = Z^4 \times \lambda^3 \times d \times t$

$30 = 1^4 \times \lambda^3 \times d \times 3$ can be written. $\qquad$ If we remember that $d = m/V$

$$d = 10/3^3 \ \text{gr/cm}^3$$

$30 = 1^4 \times \lambda^3 \times 10/27 \times 3$

$30 \times 27 = 1 \times \lambda^3 \times 10 \times 3$

$27 = \lambda^3$ and from here $\lambda = 3$

2. Let the absorption of a high-energy 2λ wavelength hydrogen photon in a 3 cm thick tissue with density d be A1, and the absorption of a high-energy λ wavelength helium photon in a 9 cm thick tissue with the same density d be A2. State the relationship between them.
   (H = 1, He =2).

   With the help of $A = Z^4 \times \lambda^3 \times d \times t$ relation,

   $A_1 = 1^4 \times (2\lambda)^3 \times 3 \times d$

   $A_2 = 2^4 \times \lambda^3 \times 9 \times d$

   If we compare these two values to each other, $A_1/A_2 = 1 \times 8\lambda^3/16 \times \lambda^3 \times 3$

   $A_1/A_2 = 1/6$

   $A_2 = 6\,A_1$

3. An electromagnetic wave with a wavelength of $1\,A^0$ collides with an electron and interacts in the form of Compton scattering. After the interaction, calculate the energy (eV) of the electromagnetic wave deflected by an angle of 90^0 and what percentage of its initial energy it loses.

   $(m_e = 9.11 \times 10^{-28})$ $(h = 6.6 \times 10^{-27}$ erg.sn$)$ $(c = 3 \times 10^{10}$ cm/s$)$

   $\Delta\lambda = \lambda' - \lambda = (2\,h/mc)\sin^2 q/2$ veya $\Delta\lambda = \lambda' - \lambda = (h/mc)\,(1 - \cos\theta)$

4.2   X-Ray Imaging Techniques

It is the oldest and most frequently used imaging technique and is preferred due to its anatomical detail and spatial resolution. In cases where radiation is risky and the X-ray method cannot distinguish, other imaging techniques are used. The method evaluates the absorption of X-ray photons and reveals the density-anatomical junction and thickness differences of the tissues. The concept that includes these parameter differences between the examined structure and its surroundings is called contrast. In radiography, the image is in the form of a two-dimensional projection of three-dimensional organs. In radiology, X-rays obtained by the collision of accelerated electrons with a metal surface under a voltage of 50–100 kV are generally used. The discrimination capabilities of these rays are quite high. Especially bone tissue, foreign metal objects, and air-filled cavities can be easily separated from the soft tissues around them. It is of great importance that X-rays are incident on some fluorescent substances in the formation of the image. Some of these substances are calcium tungstate, zinc cadmium sulfide, and cesium iodide. As a result, some visible shapes are obtained. The image is obtained by X-rays creating invisible changes (latent image) especially in AgBr crystals and by chemical processes making them visible. Applied chemical processes:

(a) **Development (first bath):** Various reducing agents such as pyrogallol and hydroquinone are used. There are opaque Ag particles in the sensitive AgBr structures. If X-rays are not sufficiently visible, there are areas observed as yellow spots. If these are not removed and exposed to light, they darken and distort the image. It is a slightly basic bath.

(b) **Fixation (second bath):** Sodium hyposulphite solution is used. It forms water-soluble salts with AgBr and ensures that the film is devoid of AgBr. Black areas called grain, which are more dense in areas that have received more X-rays, are formed. This bath also contains some substances that strengthen the gelatin structure. It is a bath in an acidic environment.

In addition to chemical processes, digital imaging systems are now frequently utilized:

(a) **Computed radiography**: In computed radiography (CR), an imaging plate made of a photostimulable phosphor is placed inside a lightproof cassette and exposed to the patient's X-ray image. Subsequently, the plate is removed from the cassette and read while being protected from ambient light. After the reading process, the image is erased, and the plate is reinserted into the cassette for reuse. CR belongs to a class of systems known as reusable plate technologies and directly replaces screen-film systems. Although the image quality of CR is lower than that of digital radiography (DR) and it requires a longer exposure time to obtain acceptable images, there are currently no alternative technologies available for reusable plates.

(b) **Digital radiography:** Digital radiography (DR) is a modern medical imaging technique that converts X-rays directly into digital images using digital detectors instead of traditional film-based radiography. Unlike CR systems, DR employs flat panel detectors or other digital receptors instead of phosphor plates, allowing images to be instantly transferred to and processed on a computer screen. The key digital technology enabling significant advancements in medical X-ray applications is the flat panel active matrix array, originally developed for laptop displays. This technology consists of large-area integrated circuits containing an active switching device—a thin-film transistor—and is referred to as active matrix arrays. These arrays function as a self-scanned readout system, in which the image generated in a particular plane is read out within the same plane. They are manufactured by depositing hydrogenated amorphous silicon onto a thin (~0.7 mm) glass substrate, facilitating large-area production.

4.2.1 Comparisons of Digital and Analog Systems

Advantages of digital radiography over analog systems include:

- Image Quality and Radiation Dose Benefits:
  - Generally requires a lower radiation dose for image acquisition
  - Provides higher resolution in certain applications
  - Offers a greater dynamic range, enhancing image contrast and detail visibility

- Operational and Practical Advantages:
  - Eliminates the need for manual handling and transportation of cassettes
  - Enables immediate assessment of image quality and patient positioning
  - Facilitates the efficient transmission and sharing of digital images

4.3 Improvement of the Image

4.3.1 Bucky Diaphragm

It is a mechanism made up of lead plates with a thickness of 0.002–0.005 cm and is used to obtain a better image. It prevents the second photon that occurs during the Compton event, which is one of the absorption mechanisms of X-rays, from blurring the image.

4.3.2 Contrast Agents

The most important substance used for this purpose in our medicine is barium sulfate. It is used especially in dog practice to observe the digestive system more easily and to determine whether there is a foreign body or a pathological condition change.

Fluoroscopy A technique for reflecting X-rays onto a fluorescent screen and examining the image on the screen at that time.

Angiography A type of imaging performed by injecting contrast material into the vessels to visualize the boundaries of the vessels that feed organs such as the heart and brain.

Other Imaging Techniques

5

Abstract

In this section, a comprehensive evaluation is made in terms of other imaging techniques and information is provided about laser beam and laser therapy. Thermography's place in medicine is emphasized and information is provided about sound waves and especially ultrasonography and echocardiography are examined and details are given about the piezoelectric principle. Electroencephalography, which is increasingly important in medicine, and the resulting EEG waves are evaluated by examining the montage axes. Electrocardiography, which is important in terms of heart examinations, is explained by defining the Einthoven triangle; information is given about the heart's stimulus system and montage types are specified. Electrocardiogram and the meanings of the differences are explained, and basically, phonocardiography and vector electrocardiography were explained. At the end of each section, the subject is detailed with explanatory questions.

Keywords

Thermography · Echocardiography · Electroencephalography · Einthoven triangle

5.1 Electroencephalography

Electroencephalography is an imaging technique that uses electrodes placed around the skull to detect electrical activity in the brain, which is the central control organ, and detects electrical potential changes. Frequency and amplitude have special importance in the evaluation. The frequency of EEG waves increases as the brain's activity level increases while their amplitude decreases. EEG waves are divided into five according to their frequencies:

1. Delta (δ) waves (0.5–4 Hertz)
2. Theta (θ) waves (4–8 Hertz)
3. Alpha (α) waves (8–13 Hertz)
4. Beta (β) waves (13–30 Hertz)
5. Gamma (γ) waves (30–100 Hertz)

Delta waves are the slowest but highest amplitude brain waves. They are typically found during deep sleep, with frequencies ranging from 0.5 to 4 Hz and variable amplitudes. They may be generalized or focal. If recorded during wakefulness, they indicate abnormality and widespread cortical dysfunction. Delta waves may be rhythmic or arrhythmic.

Theta waves are also classified as "slow." They are waveforms observed at a frequency of 4–8 Hz and amplitudes greater than 20 Mv. They are recorded in situations of extreme relaxation and tendency to sleep.

Alpha waves are located in the middle of the brain wave spectrum. These waves are observed with a frequency of 8–13 Hz and an amplitude of 30–50 Mv when the brain is not directly focused on a task, when relaxed, not in a state requiring concentration, when awake, and with eyes closed.

Beta waves are a high-frequency (13–30 Hz) and generally low-amplitude (5–30 Mv) rhythm associated with focus. If they constitute more than 20% of the total recording, they can usually indicate an abnormality or drug effect. Higher beta waves are observed in awake and decision-making states.

Gamma waves are related to cognitive function. These waves are very high frequencies where the brain produces the fastest brain waves. These range from 30 Hz upward.

EEG reflects the electrical activity of cortical neurons. The EEG method allows the analysis of macroscopic electrical activity of the brain surface layer by means of electrodes on the head. EEG is a very effective examination method in determining brain functions today. It has special importance as a routine clinical diagnostic tool for various nervous system disorders. It is of great importance in the diagnosis of intracranial lesions as a supporter of the results of clinical, laboratory, radiological, and MRI examinations and as a prognosticator.

5.2   Electroencephalography Montage

The arrangement of electrode combinations used in EEG recording is called a montage. In addition to the individual discharges detected by each electrode, the association of these electrodes with each other creates a much richer data set across the entire scalp. There are two main types of display montage: bipolar and monopolar/reference.

Monopolar, or in other words the reference montage method, is a montage type in which the voltage/potential differences between each electrode and a single fixed reference electrode, usually the average of all electrodes, are compared. Bipolar montage method is a montage type consisting of electrode chains in which the potential differences between adjacent skull electrodes are measured.

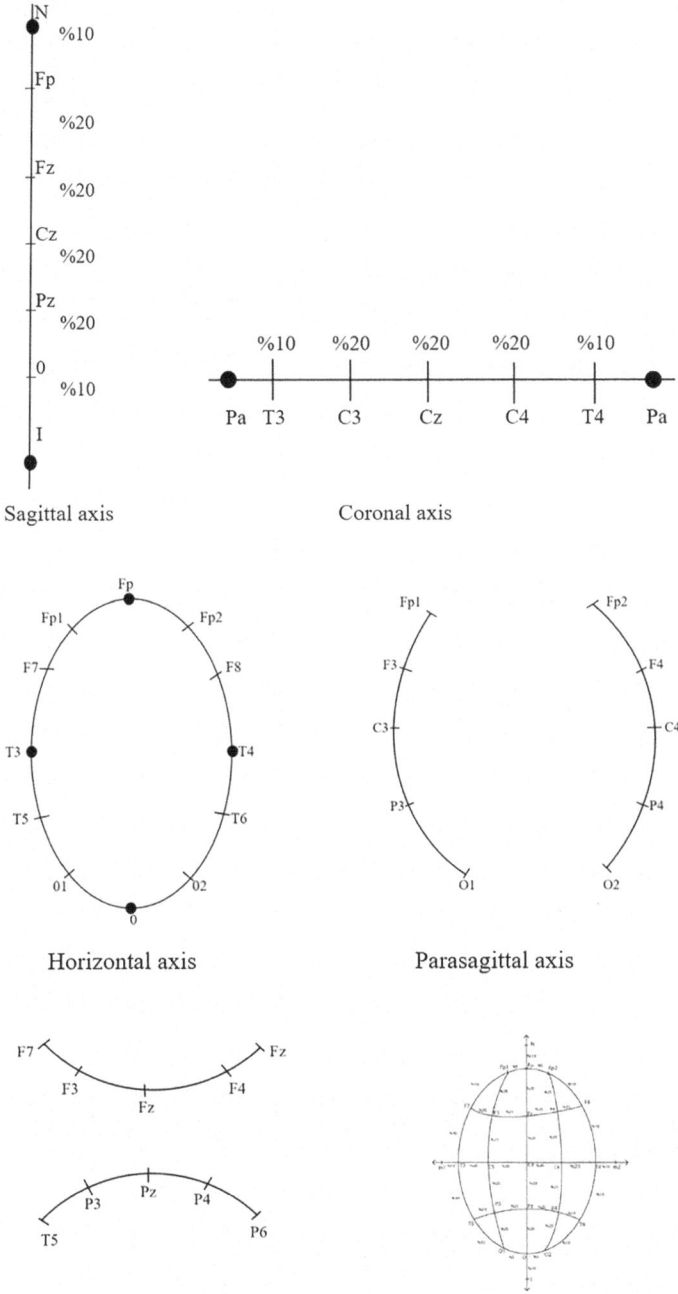

Fig. 5.1 Axes used in electroencephalography

This stage is reached after the patient is calmed down and the preparation is completed so that the electrodes can be placed according to the specified assembly technique. The electrode locations on the dog's head are given with their respective abbreviations. Each electrode point is first characterized by the initials of the areas of the brain where they are located. These abbreviations are "F" for frontal, "T" for temporal, "P" for parietal, "O" for occipital, and "C" for central. Then the right cerebral hemisphere is coded in the international system with even numbers and the left cerebral hemisphere with odd numbers. Reference points are usually indicated with the letter "R" or "Ref."

There are five axes that are used especially in human medicine and are being adapted to veterinary medicine:

1. Sagittal axis
2. Coronal axis
3. Horizontal axis
4. Parasagittal axis
5. Precoronal and postcoronal axis

The points of these axes, separately and all determined on the same skull, are seen in Fig. 5.1.

5.3   Electrooculography

Electrooculography (EOG) is a method for detecting eye movement by recording the potential resulting from hyperpolarization and depolarization between the cornea and the retina. The signal generated during the process is called an electrooculogram. The EOG value varies between 50 and 3500 μV with a frequency range of approximately 0.1–100 Hz. The EOG signal varies by approximately 20 μV for each degree of eye movement. This signal is the result of a number of factors, including eyeball rotation and movement, eyelid movement, different artifact sources such as EEG, electromyography (EMG), electrode placement, head movements, the effect of lighting, etc.

5.4   Laser (Light Amplification by the Stimulated Emission of Radiation)

Laser, an artificial light that does not exist in nature, can be defined as the amplification of light by the stimulated emission of radiation.

Properties of laser light:

1. The photons that make up the laser are in phase. The beams travel in the same direction.
2. Laser light is monochromatic, that is, single color.

3. Laser light photons have the same energy.
4. Laser light has a dipole moment.
5. Laser power density is quite high.

The temperature increases with the laser effect, van der Waal's bonds are broken, and the cell dies. In our medicine, there are different types of lasers applied, such as infrared laser (for coagulation purposes), CO_2 laser (cutting tissue, bone, closing the veins), and very thin laser beams (changes on chromosomes). It has been reported that laser, in particular, increases the phagocytic activity of leukocytes, stimulates wound healing, and has an acupuncture-like effect in pain relief.

5.5 Thermography

It is an imaging method for determining the distribution of infrared heat rays emitted from the body. Infrared ray (IR) or thermal radiation is produced by the thermal vibration of radiations with wavelengths longer than visible light but shorter than microwave rays and is between $\lambda = 750$ nm–1 mm. Infrared light intensity is directly proportional to the temperature of the tissue emitting heat and the blood flow within the tissue. Infrared ray is divided into four according to wavelength:

1. Short wavelength: 3–4 µm
2. Medium wavelength: 3–8 µm
3. Long wavelength: 8–15 µm
4. Ultra long wavelength: 15–1000 µm

Thermal imaging is a noninvasive diagnostic technique that does not require the application of an additional catheter to the patient, the administration of contrast material, or the application of ionizing radiation, allowing the operator to see changes in the patient's skin surface temperature. It is used to show physiological changes and metabolic processes by utilizing surface temperature distribution. Body surface temperature will vary depending on age, gender, weight, metabolism, topography, and the thermal effect of blood flowing through the vessels. Very small temperature differences in painful or abnormal areas of the body and infrared waves emitted from the skin surface are detected by electronic systems with detectors sensitive to 8–14 µ and converted into electrical signals. These signals are converted into colored digital images by utilizing the differences in the color scale, and as a result, a thermal image of the human body called a thermogram is obtained.

Thermal imaging is a noninvasive method used for diagnostic purposes in veterinary medicine. This method shows that the damaged tissues and organs of animals emit heat differently than normal, and the diseased area can be determined. The infrared thermal camera detects the heat and infrared radiation created by the blood flow in the skin capillaries. Thanks to thermography, an advanced technology, the body temperature information formed on the skin of animals is displayed as colored images. This method, which is especially preferred in the diagnosis of orthopedic disorders in horses, also yields successful results in research conducted on sheep, pigs, and poultry farm animals. Imaging with thermal cameras is also used in methods covering areas such as reproduction, thermal balance, animal health, and milk processing. Applications are made according to the increase and decrease in temperature in the tissue depending on the blood flow. Thermal imaging is very important in the early diagnosis of mastitis in dairy cows. The temperature change in the udder skin can be observed with thermal cameras. Thermal imaging is a useful method in determining regional sensitivity in diseases that cause heat increase and is used in the detection of temperature changes related to clinical mastitis and other diseases in dairy cows.

5.6   Ultrasonic Imaging Techniques

Ultrasonic diagnostic methods are an imaging technique based on the principle of using special sound waves. Sound propagates in waves in solid, liquid, and gaseous media, except for space. These waves can be classified according to their frequencies as follows:

1. Infrasound: Sound waves with frequencies less than 16 cps (cycles per second-hertz).
2. Normally audible sounds: Sound waves with frequencies between 16 cps and 16,000 cps.
3. Ultrasound: Sound waves with very high frequencies (20,000 cps<). Ultrasound waves with frequencies between 1 and 10 Megahertz are generally preferred for the medical field.

5.6.1   Piezoelectric Principle

According to this principle, put forward by Pierre-Jacques Curie, substances such as quartz, which are naturally crystalline, create an electrical potential in a certain direction if subjected to a mechanical force. Similarly, it is clear that mechanical energy will be generated if an electrical potential is applied to certain directions of these crystals. Since quartz is not suitable for diagnosis, synthetic crystals such as barium titanate and lead zirconate have been used. In particular, a crystal structure called transducer is used in the ultrasonographic imaging technique.

Physical properties of ultrasound waves:

1. Ultrasound waves travel longitudinally in all tissues except bone at a speed of approximately 1550 m/sec (velocity), similar to a sine wave.
2. Ultrasound waves are strongly reflected (reflection).

3. Ultrasound waves are weakened in tissues (attenuation). This amount is the sum of the energy lost by absorption, scattering, and reflection. The average attenuation is given as 1 decibel (dB) for each MHz and cm.

4. Ultrasound waves encounter resistance depending on their velocity and the density of the environment they affect (acoustic impedance). It has been reported that this value is high especially in soft tissue and air or bone passages, and this value is low in homogeneous environments. The basis of ultrasonic imaging methods is the waves that come to the surfaces separating different tissues in the organism and reflect (pulse-echo methods).

5. Sound power of ultrasound waves (sound intensity): Generally, sounds with a power of less than 10 mW/cm^2 are used. If decibels are desired to be used as a unit, the logarithmic expression of the ratio between the powers of two sound waves is found.

$$1 \text{ dB} = 10 \ \log \ P_1/P_2$$

6. Ultrasound waves are reflected at the same angle as they entered the tissue. However, waves coming from boundary surfaces that do not form a right angle with the direction of the sound wave cannot be reflected. This is not the case for rough surfaces. Even if they do not form a right angle, they can be observed by scattering.

7. The image is obtained by the reflection effect of ultrasound waves. The shortest distance that provides lateral detail is called lateral resolution.

Types of ultrasonic imaging:

1. A. Mode: It is a type used to measure distance and thickness, and a one-dimensional image is obtained.

2. M. Mode: It is a one-dimensional type that allows echoes reflected from tissues to be seen as bright spots.

3. B. Mode: Echoes are detected as light dots and a two-dimensional image is obtained. With this method, organ boundaries are displayed as gray sections.

Sonographic scanners are classified as parallel scan, sector scan, and compound scan.

A sonographic section can be depicted by moving the transducer over a certain plane, viewing the echoes as light dots on a screen, and summarizing these dots.

The imaging method related to the heart, where the pulse-echo principles of moving structures are applied, is called echocardiography. Especially since the heart is completely mobile, M. Mode echocardiography is applied. The sound emitted from the transducer is reflected in the tissues, the waves are displayed on the monitor, and, if desired, a record is taken on a paper. In our medicine, it has found a lot of use in various atrium and ventricular dilatations, hypertrophies, and heart valve errors related to running performance observed in dogs, racehorses, and foals.

In the Doppler method, a dynamic image was attempted to be obtained by considering the changes in the frequency of continuous ultrasound reflected from moving surfaces. Three types of Doppler are used: continuous, pulsed, and color. It has provided great ease of use, especially in the examination of the fetal heart and respiration, determination of the peak blood velocities between chambers, determination of the blood flow rate in the vessels, and detection of heart valve errors. Here, the blood coming toward the transducer is seen as red, and the blood moving away is seen as blue. Simple Doppler systems, such as small handheld devices, frequently utilize continuous wave (CW) Doppler. A CW Doppler transducer consists of two adjacent piezoelectric elements positioned at a slight angle to each other. The transmitter emits a continuous sinusoidal wave in the form of $\cos(\omega_0 t)$, while the receiver detects echoes returning from the region of overlap between transmitter and receiver beams. The primary limitation of continuous wave (CW) Doppler is its lack of spatial resolution, which arises from the extensive overlap between the transmitter and receiver beams. This issue is addressed by pulsed wave (PW) Doppler, which transmits a sequence of short pulses—similar to those used in imaging—rather than a continuous sinusoidal wave.

Questions 17

1. If a sound wave has a wavelength of $\lambda = 0.5$ Å and a frequency of $f = 3s^{-1}$ (3 Hz), calculate its velocity.
2. It has been determined that 15% of the ultrasonic wave coming to any tissue is scattered, 20% is spread, and 10% is absorbed.
   (a) What is the percentage of the attenuated part?
   (b) What percentage of the tissue is affected by the wave?
3. The intensity of a sound wave is given as 1000 W/cm^2, and the intensity of another sound wave is given as 10 W/cm^2. Calculate the intensity of the sound wave in decibels.

5.7   Anatomical Structure of the Heart

If we compare the heart to a box with four compartments, the upper chambers are called atrium and the lower chambers are called ventricles. The oxygen-rich blood coming out of the left ventricle of the heart (ventricles sinister) through the aorta, after circulating throughout the body, and is collected by the veins and brought to the right atrium (atrium dextrum) by the cranial vena cavae (caudalis) is known as the aortic circulation (greater circulation). The dirty blood coming out of the right ventricle of the heart (ventricles dexter) through the right ostium atrioventriculare, is carried to the lungs, and after the oxygen and carbon dioxide transformation in the capillaries in the alveoli, returns to the left atrium (atrium sinistrum) by the pulmonary vein is called the pulmonary circulation (small circulation). Between the atrium sinistrum and the ventricles sinistrum, there is the atrioventricularis sinistra (bicuspidal valve); between the atrium dextrum and the ventricles dextrum, there is the atrioventricularis dextra (tricuspidal valve); and at the mouth of the aorta and pulmonary artery, there are semilunar valves.

5.7.1 Heart's Stimulus and Conduction System

There are two important nerves that provide innervation to the heart. The first of these, N. vagus, slows down the heart's work, while N. accelerantes, on the contrary, speeds up the heart's work.

Another important system that affects the heart's functioning is an internal system that starts from the sinoatrial (S-A) (Keith-Flack) node, where the right atrium meets the cranial v. cava, and spreads to the entire heart. An impulse created spontaneously in the sinoatrial node stimulates the atrium (depolarization), and this stimulus is transmitted to the atrioventricular (A-V) (Ashoff-Tawara) node located at the lower part of the atrium wall by neuromuscular fibers. As a result of this event, the atrium muscle cells enter a resting state (polarization). Then the stimulus is transmitted to the ventricular muscles by the bundles of His (chordae tendineae), which are divided into two on the right and left. The ends of the bundles of His have a widespread network-like structure (Purkinje network). Purkinje cells are cells that do not have the ability to contract but produce electrical stimulation. These cells are specialized to create action potentials.

In the cell membrane theory, during the resting period, there are K^+ ions inside the cell and Na^+ and Cl^- ions outside. There is no electrical potential between the two sides of the cell membrane (polarization). When the cell is excited, the Na^+ permeation of the cell membrane increases. The inside of the cell is filled with more positive charge than the outside (depolarization).

The action currents that occur during the heart's work are spread throughout the body. Since body fluids are conductive, the recording process defined as the entry of the current from the two poles (electrodes attached to the body surface) into a galvanometer is called electrocardiography. The records obtained as a result of this process are called electrocardiogram. The basic principle is that the depolarization waves spread from different parts of the ventricles create an electric dipole. There can be many points on the body that can be taken in this way. In order to eliminate differences in cardiology, some standard points have been accepted. Here, derivation is a term that expresses the electrode systems used and the areas where they are placed. Derivations are divided into standard extremity derivations, augmented extremity derivations, and chest derivations. Derivations used in humans:

1. Standard limb derivations (bipolar):
   (a) Negative electrode on the right arm, positive electrode on the left arm: D_1 or Lead I
   (b) Negative electrode on the right arm, positive electrode on the left leg: D_2 or Lead II
   (c) Negative electrode on the left arm, positive electrode on the left leg: D_3 or Lead III
   It is especially used in the examination of heart rhythm disorders.
2. Augmented limb leads (unipolar):
   A. Positive electrode on the right arm: aVR
   B. Positive electrode on the left arm: aVL
   C. Positive electrode on the left leg: aVF

It is especially important in the diagnosis of infarction.
3. Chest leads:
   V_1, V_2, V_3, V_4, V_5, and V_6

It is especially important in the diagnosis of cardiac hypertrophy and heart block.

5.7.2 Einthoven Triangle

The heart is thought of as an electromotive force placed in the middle of a triangle whose base is upward and apex is downward in the chest cavity. The base corners of the triangle show the electrical point where the right and left arms meet the chest, and the apex of the triangle shows the electrical point where the left leg meets the chest. The sides of this triangle show the derivation axes. In other words, the base of the triangle gives derivation I. If the positive end of this axis is accepted as 0o and parallels are drawn from its midpoint to the other two edges, derivation axes are obtained at different angles.

5.7.3 Einthoven Law

If the potential in any two of the standard limb leads is known, the potential in the other lead can be found with the help of the

$$I + III = II$$

relation.

In animals, the standard and augmented extremity derivations are the same except for the horse. When taking the electrode, the animal must be completely still and any stimuli from the environment must be reduced as much as possible. While standing electrodes are taken from horses and cattle, dogs are taken both standing and lying on their right side. Before starting the ECG recording, calibration (adjustment) is made to obtain a 1 cm deviation with a voltage of 1 mV. The speed of the paper is 25 mm/s, and 1 mm represents 40 ms.

5.7.4 Electrocardiogram

A normal trace consists of three parts (Fig. 5.2).

(a) **P wave:** It shows the spread of the impulse from the sinoatrial node to the atria, and therefore the depolarization of the atria. The P wave is positive in horses, cattle, and dogs, and sometimes this wave can be notched. The P-Q interval represents the time from the beginning of depolarization of the atria to the depolarization of the ventricles.

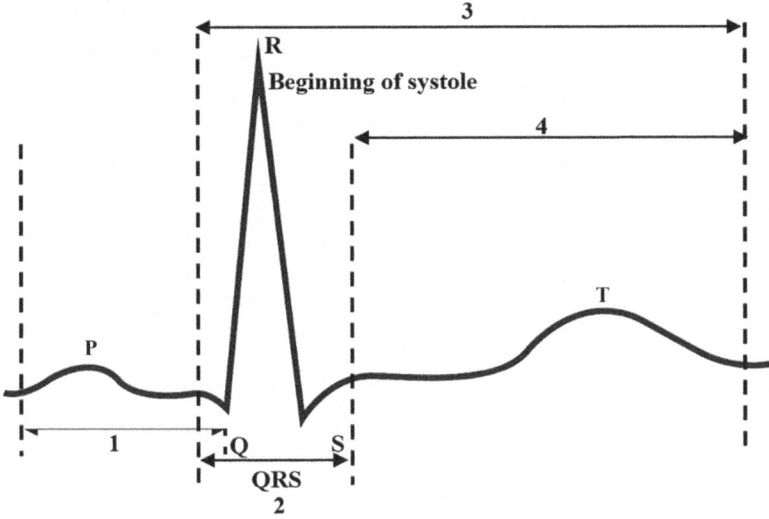

Fig. 5.2 Normal ECG trace

(b) **QRS complex:** It shows the spread of the stimulus in the ventricles, that is, the depolarization of the ventricles. The Q wave is the small, first negative wave that occurs after the P. The deepening of the Q wave indicates a past myocardial infarction. R is a distinct positive wave. S is the second negative wave of the complex. The S-T segment is the part from the end of the QRS complex to the beginning of T. The fact that this part is above or below the isoelectric line should be evaluated for conditions such as pericarditis, myocardial ischemia, and hypokalemia.

(c) **T wave:** It indicates that the ventricles are repolarized, that is, they are returning to a resting state. It occurs 0.25–0.30 s after depolarization.

Repolarization of the atria is not observed on the ECG because this event occurs during depolarization of the ventricles.

5.7.5   Artifacts

These are changes that occur on the ECG curve due to external factors not related to the heart, such as electrical interference, incorrect techniques, and application errors, without any pathological change.

1. Muscle tremors: The patient who will receive an ECG is overly excited, scared or very nervous.

2. Curvature of the isoelectric line: Movement of the patient on whom the ECG will be taken, shaking of the table on which animal lies, very frequent and strong breathing, the electrodes used not being clean, not using gel, and rumination in ruminants.
3. Alternating current interference: It is related to ground line connection problems.

5.7.6 Phonocardiography

The recording of heart sounds with the help of special microphones, filtering, and then transferring these sounds to tapes is called phonocardiography. Phonocardiogram is obtained by transferring this data to paper. Congenital anomalies, heart valve defects, and arrhythmias can be detected with this method.

5.8   Vector Electrocardiography

It is a method that can be defined as the determination of the curve drawn in three-dimensional space by the electric dipole in the heart, which occurs during the depolarization of the atria and ventricles and the repolarization of the ventricles. Although it is used in horses, it is not a very practical method.

Questions 18

1. If the first lead of a dog brought to the clinic to detect heart rhythm disorders is measured as 5 mV and the third lead is measured as 14 mV from the standard extremity leads, calculate the potential change in the second lead by writing the necessary formula.
2. If the first lead of a dog brought to the clinic to detect heart rhythm disorders is measured as 3 mV and the second lead is measured as 14 mV from the standard extremity leads, calculate the potential change in the third lead by writing the necessary formula.
3. Name the compartments and vessels of the heart by showing them on a figure. (Show the pulmonary circulation with a blue pen and the body circulation with a red pen.)
4. Draw the conduction system on the heart schematically and name it.

Resources

Aytuğ CN (1997) Hormone therapy in veterinary medicine. Taç Ofset Printing House, İstanbul

Barutçu B (1993) Biophysics practical notes. I.U. Cerrahpaşa Medical Faculty Publications, Istanbul

Başar G (1998) Physics III lecture notes. IU Faculty of Science, Istanbul University, Istanbul

Bilgiç B, İskefli O, Pugliese M, Or ME (2025) Evaluation of Cardiomegaly in Dogs Using the Manubrium Heart Score Method and Determination of Its Diagnostic Accuracy in Comparison with the Vertebral Heart Score. Vet Sci 12(7):619

Bilgiç B, Tarhan D, Or ME (2024) The effects of different treatments on serum trace element levels in dogs with heart failure. Animals 14(23):3390

Bilgiç B, Tarhan D, Ekiz B, Ercan AM, Or ME (2023) The effect of pimobendan and enalapril use on blood serum trace element levels in dogs with myxomatous mitral valve disease. Journal of Trace Elements and Minerals 4:100065

Biran L, Yarız E (1991) General mathematics, 2nd edn. Marmara Univ. Tech. Ed. Department, Istanbul

Bushberg JT, Seibert JA, Leidholdt EM, Boone JM (2012) Diagnostic radiology. In: The essential physics of medical imaging, 3rd edn. Lippincott Williams & Wilkins, Philadelphia, pp 169–209

Cafer S, Gönül R, Or ME (2025) Kedi ve Köpek Dermatolojisinde Kriyoterapi Uygulamaları. Journal of The Faculty of Veterinary Medicine Erciyes University 22(1):51–57

Çevik Ç, Kaya E, Turanoğlu B, Tarhan D, Bilgiç B, Ercan A, Or E (2023) Köpeklerde Elektroensefalografi ve Elektrookülografi. Dicle University Journal of Faculty of Veterinary Medicine 16(1):51–58

Cokyuksel O, Ober A, Camuscu S, Numan F (1987) Introduction to X-ray physics. IU Cerrahpasa Medical Faculty Publications, Istanbul

Dance DR, Christofides S, Maidment ADA, McLean ID, Ng KH (2014) Special topics in radiography. In: Diagnostic radiology physics: a handbook for teachers and students, 1st edn. International Atomic Energy Agency, Vienna, pp 241–247

Güner Z (1982) Physics for medical and biology students I, 3rd edn, issue: 426. Ankara University Medical Faculty Publications, Ankara

Ilgar L (1996) Educational management. Beta Publications, Istanbul

İmren HY (1994) Introduction to veterinary internal medicine. Medisan Publications, Ankara

Kraft W, Dürr UM (1997) Clinical laboratory diagnostics in veterinary medicine. Schattauer Verlag, Schattauer

Marşoğlu A (1988) General astronomy II. I.U. Faculty of Science Publications, Istanbul

Özkan MT (1994) Binary stars and high energy astrophysics lecture notes. IU Science Faculty, Istanbul

Pehlivan F (1997) Biophysics. Hacettepe-Taş Kitapçılık Ltd. Şti, Ankara

M. E. Or, *Medical Physics for Veterinary and Related Studies: An Introductory Textbook on Mathematical and Physical Principles*,
https://doi.org/10.1007/978-3-031-97355-0

Pugliese M, Napoli E, La Maestra R, Or ME, Bilgiç B, Previti A, ... Passantino A (2023) Cardiac troponin I and electrocardiographic evaluation in hospitalized cats with systemic inflammatory response syndrome. Vet Sci 10:570

Saygaç T (1999) Spectroscopy and stellar science lecture notes. IU Science Faculty, Istanbul

Taylor JR, Zafaritos C (1996) Modern physics in physics and engineering. Güven Publications, Istanbul

Tektunalı HG (1990) Introduction to astrophysics I. I.U. Faculty of Science Publications, Istanbul

Terzioğlu M, Yiğit G, Oruç T (1993) Physiology volume II. University Cerrahpaşa Medical Faculty Publications, Istanbul

Turgut K (1997) Veterinary clinical laboratory diagnosis. Selcuk University Veterinary Faculty, Konya

Yıldırım C (1988) Mathematical thinking, Series: 87. Remzi Bookstore, İstanbul

The manufacturer's authorised representative in the EU is Springer
Nature Customer Service Centre GmbH, Europaplatz 3, 69115 Heidelberg,
Germany. If you have any concerns regarding our products, please
contact ProductSafety@springernature.com

Printed and bound by CPI Group (UK) Ltd, Croydon, CR0 4YY

01/05/2026

02100946-0001